THE SECRET

OF SUCCESS

The Double Helix of Formal and Informal Structures in an R&D Laboratory

POLLY S. RIZOVA

STANFORD BUSINESS BOOKS
An Imprint of Stanford University Press
Stanford, California

Stanford University Press
Stanford, California

Library of Congress Cataloging-in-Publication Data

Rizova, Polly S.
 The secret of success : the double helix of formal and informal structures in an
R&D laboratory / Polly S. Rizova.
    p. cm.
 Includes bibliographical references and index.
 ISBN 978-0-8047-5570-2 (cloth : alk. paper)
 1. Technological innovations--Case studies. 2. Business networks--Case stud-
ies. 3. Social capital (Sociology)--Economic aspects--Case studies. 4. Knowledge
management--Case studies. 5. Research, Industrial--Case studies. I. Title. II.
Title: Formal and informal structures in an R&D laboratory.

HD45.R519 2007
658.5'7--dc22

                          2007020911

Typeset by Bruce Lundquist in 10.5 / 15 Minion

Special discounts for bulk quantities of Stanford Business Books are available to
corporations, professional associations, and other organizations. For details and
discount information, contact the special sales department of Stanford University
Press. Tel: (650) 736-1783, Fax: (650) 736-1784

To Nevena Radoynovska, for whom I wrote this book,

and to Ivan Kamenov, because of whom I did.

TABLE OF CONTENTS

FIGURES

TABLES

ABBREVIATIONS

R&D Research and Development

F = High degree of formal communication

f = Low degree of formal communication

C = High degree of corporate support for the project

c = Low degree of corporate support for the project

T = A person who occupies a highly central position in the technical-advice network is on the project

t = No person who occupies a highly central position in the technical-advice network is on the project

O = A person who occupies a highly central position in the organizational-advice network is on the project

o = No person who occupies a highly central position in the organizational-advice network is on the project

S = Project with a "high success" outcome

s = Project with a "low success" outcome

ACKNOWLEDGMENTS

I am genuinely and profoundly grateful to many people for their intellectual and moral support throughout the writing of this book. First and foremost my gratitude goes to John Stone. His relentless encouragement and support, along with a close reading of the manuscript and an uplifting sense of humor, not only gave me the confidence to dream of reaching this milestone but made my writing experience rewarding.

This project would have never been conceived if it were not for Dr. John B. Bush Jr., vice president of Gillette's Corporate Research and Development (CRD). I owe him particular thanks for providing me with the once-in-a-lifetime opportunity to work with him at CRD in Boston, for the tremendously enriching three-year experience that this has been, and, above all, for igniting the passion in me to understand the complex, yet simple world of the people who have dedicated themselves to technological innovation. I am obliged to Peter L. Berger for *inviting me to sociology* and for his sophisticated and witty approach to unraveling social reality; without it, my understanding of the entangled nature of the technical and social sides of invention and innovation would have been distinctly limited. Special thanks to Stanislav Dobrev for the many constructive comments and suggestions that he offered and equally so for his encouragement, friendship, and collegiality. I am very grateful to Steven Vallas for his most helpful and candid discussion on how to frame the argument and enliven its presentation. For their feedback on various versions, portions, and aspects of this work, I wish to thank Gautam Ahuja,

Karl Aquino, Emily Barman, Steven Borgatti, Detelin Elenkov, Alden Hayashi, Satish Nambisan, Raja Roy, Lynda Aiman-Smith, Annemette Sorensen, Olav Sorenson, Bill Stevenson, David Waguespack, and the anonymous reviewers. Many thanks as well go to Jay Corrin and Linda Wells for their continual support and encouragement, and to my colleagues June Grasso, Shelley Hawks, Stephanie Kermes, Michael Kort, Susan Lee, John Mackey, Kathleen Martin, John McGrath, Edward Rafferty, Barbara Storella, Bill Tilchin, Ben Varat, and Tom Whalen for embodying the essence of collegiality.

I am most indebted to my respondents—the engineers, scientists, technicians, and managers—who have been extremely kind and generous with their time. It is to state the obvious that without them this book would have been simply impossible. Unfortunately, for reasons of anonymity, I am not at liberty to name the men and women at Global East who taught me so much about the human need for expression. Without the financial support of the National Science Foundation (MOTI-9714058), however, I would not have found out about them and their work; it has my enduring gratitude.

At Stanford University Press, I would like to thank Martha Cooley, the Economics and Business editor, and Alan Harvey, Emily Smith, and Jared Smith for their enthusiastic support for this project and their assistance in turning the manuscript into a book.

My deepest gratitude goes to my parents, who taught me never to give up and to love humankind—an impossible combination if ever there was one! They continue to give me reasons to strive and to achieve, to look for and to find meaning in my work. I cannot imagine my world without them. Finally, in the process of completing this project, I have accrued debts, yet again, to two people I do not quite know whether I will be able to repay—my daughter Nevena and my husband Miro. Their unconditional love and unwavering support through testing times, their nobility, and their genuine belief in me left no other choice but to persevere, so as not to disappoint them and to give them a reason to be proud of me, just as I am proud of both of them.

Boston, 2006

THE SECRET OF SUCCESS

1

TECHNOLOGICAL INNOVATION TODAY: OLD WINE IN NEW BOTTLES?

"In other words, 21st-century organisations are not fit for
21st-century workers."
Hindle 2006:4

Few would argue against the proposition that successful innovation lies at the heart of today's knowledge economy. Knowledge is held to be the key to economic growth in all societies, particularly within the advanced industrial states (Brint 2001; Powell and Snellman 2004). In 2006, the European Union set itself the objective of becoming "the most competitive and dynamic knowledge-based economy in the world," and the Lisbon Council Policy Brief followed rapidly with a study on "defining the role of human capital in economic growth." Part of this involved the construction of The European Human Capital Index, with the express intention of using the findings to generate policies that would lead to sustained knowledge creation and innovation across the continent (Ederer 2006). On the other side of the Atlantic, in its latest "research announcement," the IBM Center for Innovation unveiled the "Topics of Special Interest 2006–2007." Not surprisingly, key among those topics are "human capital management" and new ways of organizing for solving routine and complex problems. Companies are openly engaging in a fierce battle for talent, knowledge, and brainpower, and they are not hesitant to pay whatever price this new global market demands so long as it increases the likelihood of successful innovation. Segal reports that "In 2002, the U.S. R&D total [spending] exceeded that of Canada, France, Germany, Italy, Japan, and the United Kingdom combined . . ." and in 2004 the total expenditures on R&D in the United States were expected to reach $290 billion (2004: 3). Between 1996 and

2002, the pharmaceutical industry alone doubled its spending on R&D to $32 billion. Yet during the same period of time, the number of new drugs brought to the market fell from fifty-seven to seventeen (Canner and Mass 2005: 17). Some estimates suggest that, on average, as many as 80 percent of the innovative projects in North American companies fail either entirely or in part. Furthermore, these results are not confined to the United States. An analysis of fifty innovative projects in Dutch companies found that only one in five projects turned out to be viable (Carr 1996; Cozijnsen, Vrakking, and van Ijzerloo 2000), and this disheartening trend has not changed since 1980 (Page 1993).

This book is yet another attempt to offer a fresh perspective on how success in projects in knowledge-intensive organizations can be promoted given the contemporary challenges facing these organizations. While building on the vast body of rich theoretical and empirical insights generated by previous research, the book bridges the literature on innovation; social networks; and projects, teams, and small-group studies.

The recent thinking on innovation stresses two central features relevant to the innovative process: "first, that [it] involves the coordination and integration of specialized knowledge and, second, that it requires learning in conditions of uncertainty" (Castellacci, Grodal, Mendonca, and Wibe 2005: 94). These two features are the forces behind the fast-changing and highly uncertain environment in which knowledge-based organizations operate. As a result of the unprecedented pace at which technical and scientific knowledge has been generated in the past decade, the degree of specialization and fragmentation has increased too. General biomedical knowledge, for instance, doubles every twelve months. There are now four different databases that store three billion bytes of information in that field alone. Such trends impose substantial demands on companies to organize their laboratories and teams in ways that maximize the acquisition, targeted sharing, and utilization of the knowledge that is critical to the survival of the organization. Knowledge, however, resides in the minds of individuals, as well as being embodied in various routines, relations of power, and systems of meaning that people resort to in their daily work. As such, it does not easily lend itself to codification and, consequently, its retrieval and transfer is not a small and trivial matter. Another factor that complicates the ability of both individuals and organizations to process knowl-

edge has to do with the fact that knowledge is dynamic and therefore elusive as it evolves over time. The implications of this only increase in number with the advent of a global economy that creates R&D settings that are geographically dispersed, and, at the same time, functionally organized. One plausible consequence of these trends is that innovation can originate from any place in the company in geographical, functional, and hierarchical terms, given proper support. In consequence, to manage innovation is to perpetually attempt to manage both knowledge and uncertainty.

The question then is what can be done to reduce the uncertainty and arrange for the coordination and integration of highly specialized and widely dispersed knowledge? That the structure of a system has the ability to encourage certain patterns of interaction as well as outcomes while constraining others is one of the main tenets of social science. In view of this, Fagerberg's suggestion that we need to explore the answers to two main questions if we are to attempt to unravel the complexities surrounding the management of technology and innovation seems promising. They are "Is the potential for communication and interaction through existing linkages sufficiently exploited? Are there potential linkages within the system that might be profitably established?" (2005: 13). These questions, he states, apply to both systems and social networks. To understand how technological outcomes can be shaped, then, one must focus on the manner in which unit structures—both formal and social networks—affect the ability of an organization to effectively locate, access, and transfer critical knowledge in conditions of uncertainty. Do these structures act separately or in interaction? If the latter, how? In what manner can an R&D organization make best use of its human and social capital to achieve technological success? Can social relations be managed to the benefit of technological success? Is it possible to design technical projects that draw on the advantages of the formal and social network structures while avoiding their downsides? These are the questions that lie at the center of this book.

Despite groundshaking changes in the nature of work and authority patterns (Barley and Kunda 2004; Brint 2001; Kleinman and Vallas 2001; Vallas 1999), experts acknowledge that "today's big companies do very little to enhance the productivity of their professionals. In fact, their vertically oriented organizational structures, retrofitted with ad hoc and matrix overlays, nearly

always make professional work more complex and inefficient" (Hindle 2006: 4). Twenty-first-century organizations may need to devise yet another design form that reflects and accommodates the nature of the profound changes taking place in knowledge enterprises. This new design is likely not only to entail changes in the way in which positions and job descriptions are conceived but to require a considerable adjustment in the behaviors and attitudes of their occupants. For instance, most specialists would agree that the benefits of designing a decision-making process in a manner that individual team members can and will contribute to are practically immeasurable. This will involve the creation and maintenance of a different organization though; one in which there are new and meaningful ways of rewarding knowledge workers, particularly given the trend toward flat structures and the increased reliance on teams and project-based work;[1] one in which flexibility and perpetual change in the composition of the task environments may become the only constant. In light of this, a major emerging issue in contemporary R&D organizations is not just how to design formal channels to best navigate knowledge and information but, equally important, how to guide the informal relations between managers, scientists, engineers, and technicians with whom a large part of the knowledge and expertise resides.

This book is based on the premise that, although it is the individuals in possession of specific talents and skills who come up with novel ideas and connect the dots between remote and seemingly unrelated bits of knowledge and information, technological innovation is a process that takes place in a social setting. Hence, my interest is in the factors that enhance and constrain this social process. In particular, I am concerned with the conditions that create social dynamics that enable individual actors in groups to access and recombine knowledge in novel ways, so that the organization can then bring that knowledge to fruition. To this end, I pay special attention to the role of formal and social network structures as conduits of knowledge and information exchange as well as channels of decision making. My specific focus is on R&D projects and teams. Such teams, of course, do not exist in a vacuum. They are embedded in an R&D organization, a company, and, by extension, an industry (Ancona 1990). With this in mind, I examine the technical projects as part of their immediate organizational context, culture, and processes. My investiga-

tion is informed by structuralist approaches in the research on technological innovation, project management, and team and small-group studies.

RESEARCH ON TECHNOLOGICAL INNOVATION

The topic of technological innovation has been fascinating generations of scholars, managers, and policymakers ever since Schumpeter's influential *Theory of Economic Development* appeared in English in 1934. In it, he argued that the survival of firms, as well as that of society, is dependent upon their ability to continuously find new uses for existing resources and to recombine them in novel ways.

Information channels are the basis for any social action (Coleman 1988). Accordingly, technological innovation has been understood as a process, the result of which is the successful transfer of an idea into a new product or process that has social and market value (Allen 1977; Kerssens-Van Drongelen, Weerd-Nederhof, and Fisscher 1996; Van de Ven 1986). Hence, technological innovation is the result of information- and knowledge-processing activities that take place in an organizational context. These go beyond the generation of a creative idea in the minds of single individuals and proceed in stages (Ebadi and Utterback 1984). As innovations are conceived and accomplished by people within organizations, then both the individuals and the organizations are essentially information-processing units (March and Simon 1958). These units digest the scientific, technical, managerial, and contextual information acquired by and transmitted to the R&D staff through external and internal communication channels within the organizations (Allen, Tushman, and Lee 1979; Ancona and Caldwell 1992; Tushman 1978). Those organizations capable of effectively sharing knowledge are, by and large, seen as pregnant with innovative capacities. Knowledge, however, just as information, is often "sticky," as Von Hippel (1994) aptly terms it, and therefore it is shared reluctantly and spreads with difficulty.

Despite intense scrutiny and a large number of empirical studies, the process and the outcome of technological innovation remain to a great extent enigmas. The management of technological innovation appears to be more of an art form than the rationally planned and methodically executed endeavor that managers and stockholders want it to be. This notion is supported by

the inconsistent findings concerning the ability of organizations to be reliably successful at that process (Damanpour 1991; Drazin and Schoonhoven 1996). In a recent and comprehensive review of the state of the empirical investigations on innovation, Fagerberg concluded that "in spite of the large amount of research in this area during the past fifty years, we know much less about why and how innovation occurs than what it leads to." (2005: 20).

A principal problem in innovation studies is that of understanding how innovation happens and how to organize for success. That design is a primary vehicle for shaping an organization's ability to achieve its goals is hardly a new insight. Chandler identified two main aspects of design, irrespective of whether it is "formally or informally defined": "It includes, first, the lines of authority and communication between the different administrative offices and officers and, second, the information and data flow through these lines of communication and authority" (1962: 14). Much of the attention in this regard has been centered on structure and the investigation of the effect of the formal organization on the behavior and outcomes of individuals and firms (Dougherty 2001; Lam 2005).

For nearly the first six decades of the twentieth century, research on innovation proceeded in parallel with organization theory and, hence, it was preoccupied with articulating overarching general principles and discovering the "one best way to organize." It was in the 1960s when the contingency tradition developed in the work of Woodward (1958), Burns and Stalker (1961), Thompson (1967), and Lawrence and Lorsch (1967) as a reaction to this monocausal model. In this perspective, organizational structure is shaped by the nature of the technology and then, in turn, shapes the relationships between people in the work processes. Following on the assumption that organizations can be conceived as systems and subsystems, each of which has its own characteristics, the contingency approach provided us with the insight that the system and its subparts are likely to benefit from different ways of organizing. This was first articulated and empirically verified by Lawrence and Lorsch (1967) in their seminal work on differentiation and integration mechanisms used in different organizational functions (such as sales, production, marketing, and R&D) in six plastics firms. Another celebrated classic is Burns and Stalker's 1961 study of the relationship between firms' innovativeness and

organization design in twenty British electronics and rayon firms. They argued that organic (decentralized and less formalized) structures are more conducive to technological innovations, particularly to those of a radical nature, as they are better able to respond rapidly to the ever-changing environment. In contrast, mechanistic (bureaucratic and highly formalized) forms tend to be innovation-resistant.

Accordingly, the literature on the management of technological innovation has examined various structural aspects and the conditions under which they affect the ability of both individuals and organizations to make discoveries. Prominent among these are centralization, formalization, horizontal and vertical integration, and the stage of industry development. The research informed by this tradition mostly relied on large surveys and statistical methodology to infer the effect of different structural arrangements on performance. Ultimately, it produced numerous valuable insights, and, much to the dissatisfaction of the managers of technology, one broad but frustratingly vague design rule: "it all depends."

As a result, a number of design approaches have been tried in an effort to increase the likelihood of technical success. Among those are the adoption of the matrix and various team-based designs, personnel rotation and the periodic retraining of technical personnel, and the selection of individuals with superior technical skills. To these must be added the identification of creative individuals with specific personality traits and the assignment of them to key decision-making positions. However, the record of success resulting from such measures has been inconsistent. The lack of consistent findings from this period has been largely attributed to the exclusive focus on the investigation of the formal structural attributes to the neglect of the role of the human agents in the process and the relations that they enter into (Barley 1990).

This void has been filled by another structuralist approach—the social networks perspective. Its origin can be traced to Simmel's work on dyads and triads (1902), but it was Granovetter's seminal article (1985) that carved a prominent place in social research for the individuals and the social relations that they establish and maintain. This perspective, too, looks at the linkages between positions and people to explain outcomes. Unlike the traditional structuralist approach, which views performance as a function of the relationships

prescribed by an organization chart, the social networks model seeks to capture the *actual* patterns of linkages and relations. It is based on the premise that the actors' behaviors can be understood through the informal structural configurations—such as friendship, advice, and collegial networks—that they are a part of, and the positions they occupy within them. In network terms these are positions of high or low respect, high or low status, and informal power (Burt 1992). At the center of the social network analysis is an examination of the forms and content of the stable patterns people develop in their relationships, as well as the effects that these create (Tichy 1980). Power and influence, in this perspective, come from the "actors' positions in the actual patterns of interaction that define a social network rather than from their positions in the formally defined vertical and horizontal division of labor" (Ibarra 1993: 476).

The social network literature has been prolific. It is replete with empirical evidence of the advantages that the informal structures offer over the formally prescribed rules and behaviors (Burt 2000; Hansen 2002). Likewise, the effect of social networks on outcomes has been the subject of numerous investigations. Social networks have been found to help coordinate critical task interdependencies (Blau 1955; Gulati and Gargiulo 1999; Pfeffer and Salancik 1978); to ease access to information and speed up the information exchange (Granovetter 1995; Ingram and Roberts 2000); to serve as webs of idea generation and create opportunities for learning (Hage and Hollingsworth 2000; Podolny and Page 1998); to increase learning rates (Argote, Beckman, and Epple 1990); and to produce economic benefits (Uzzi 1999). Furthermore, social networks have been shown to generate social capital (Bourdieu 1986; Lin 2001). Generally, those who occupy a more central position in the social network possess greater social capital. Scholars have identified several mechanisms through which social relations create capital assets (Burt 2000; Coleman 1988; Nahapiet and Ghoshal 1998). Among those are trust and trustworthiness; the power of social norms and sanctions; expectations that obligations will be honored; and a source of identity.

Despite consistent evidence of the connection between social networks and these themes, though, the promise and pertinence of which should be obvious to the study of innovation, "relatively few studies . . . link informal ties to the innovation process. . . ." (Powell and Grodal 2005: 70). Mote, too, observed

that "[w]hile the role of social networks in scientific research and R&D is recognized, it has often been overlooked in favor of the formal structure of the research organization" (2005: 97). What is more, the vast majority of those are conducted at the interorganizational level. Here, network dynamics have been effectively employed to explain performance and the evolution of technological fields (Fleming and Sorenson 2000; Powell, Koput, and Smith-Doerr 1996). For instance, in their 1996 analyses of biotechnology firms between 1990 and 1994, Powell and his colleagues found that an industry characterized by rapid technological development and a complex and expanding knowledge base had the locus of innovation not in individual firms but in networks of firms. They described these as networks of learning, expressed through large-scale, interorganizational collaboration. In another biotechnology study on the sourcing of scientific knowledge, Liebeskind, Oliver, Zucker, and Brewer (1996) reached a similar conclusion. In a more recent longitudinal study of chemical companies, Ahuja (2000) investigated the role that three aspects of a firm's position in an industry network—direct ties, indirect ties, and structural holes—play in the organization's ability to generate innovations. He found that each structural aspect offers a distinct contribution to innovation output. Finally, Johnston and Linton conducted research on the implementation of environmental technology in eighty-three North American firms from the electronics industry. They found that "interfirm networks composed of both suppliers and competitors were significantly correlated" with the implementation of the technology (2000: 465).

Research at the level of the firm has not been as voluminous as that conducted at the interorganizational level. Those studies that focus on intraorganizational social networks have found that central positions, and the ability to connect effectively to others within the company, grant access to knowledge and as a result improve the capacity of the organization to innovate (Ancona and Caldwell 1992; Hansen 2002). Furthermore, informal structures have been shown to be a powerful mechanism in the creative process, as central network positions generate opportunities for expanding one's communication network (Allen 1977; Ibarra 1993; Tsai 2001).

While this has been very helpful, it can be argued that research at the business-unit level does not capture the intensity and richness of interpersonal

relations in groups and on projects in which collaborative work takes place. As a consequence of the smaller size and the frequency of the interactions at this level, it is only logical to expect that the stable patterns of social relations that people establish are likely to create their own distinct dynamics and, as a result, to play a role in shaping group outcomes in their own specific ways. Teams and groups, though, have received surprisingly little attention in the network literature. In recognition of this fact, Oh, Labianca, and Chung (2006) argued the case for developing a multilevel model of group social capital. Among those relevant studies is the one conducted by Hansen, Mors, and Løvås (2005), who looked at how networks in new-product development teams affect the sharing of knowledge. In another study, Reagans, Zuckerman, and McEvily (2004) compared the effectiveness of two project team staffing approaches—one that focuses on the team members' demographic characteristics and the other on members' social networks. Borgatti and Cross (2003) examined how relational characteristics influence the individual's information-seeking behavior in groups. And Rulke and Galaskiewicz (2000) studied MBA game teams to investigate the joint effect of the distribution of knowledge and the social network structure of a group on its performance. Nevertheless, few of those investigations have had at the center of their exploration the effect of social networks on the outcomes and performance of teams and groups.

One such example of applying the social network approach at the group level is the study by Sparrowe, Liden, Wayne, and Kraimer (2001) who explored the relationship between various dimensions of the social network structure and the performance of both individuals and teams in thirty-eight work groups. Interestingly enough, they found that the same network structural characteristics affect the performance of individuals and groups differently. For instance, their results show that there is a positive relationship between occupying a central position in an advice network and an individual's performance. At the group level, however, the density of advice networks was not linked to productivity, whereas network centralization was actually found to hinder performance, particularly on complex tasks.

What is even more surprising is that the research investigating the link between informal ties and the outcomes of technical projects has been, by and large, missing from the social network literature. In fact, there are a handful

of studies that have explored this connection for R&D projects and teams. My investigation joins those few (Hansen 1999; Reagans and Zuckerman 2001; Rizova 2002, 2006a; Smith-Doerr, Manev, and Rizova 2004). In a study of 120 new-product development projects, carried out by forty-one divisions of a large electronics company, Hansen found support for his hypothesis that "weak interunit ties help a project team search for useful knowledge in other sub-units but impede the transfer of complex knowledge, which tends to require a strong tie between the two parties to a transfer" (1999: 82). Research on 224 corporate R&D teams conducted by Reagans and Zuckerman (2001) demonstrated that both high network density and high network heterogeneity explained team productivity. Smith-Doerr, Manev, and Rizova (2004) subsequently revealed how managers' positions of centrality in social networks shape the social construction of the outcome of innovation projects.

All in all, current research on social networks has produced consistently valuable knowledge about how the structure of social relations affects innovation at the inter- and intraorganizational levels. From these studies, it is also clear that there is a far more sophisticated and subtle understanding of how innovations arise out of the complex interaction between social networks and social capital within R&D projects. Given the nature of the challenges that today's knowledge-based organizations face, the changing structures and authority patterns in knowledge-intensive organizations (Kleinman and Vallas 2001), and the heavy reliance on team and project-based work (Griffin 1997), the overlook of R&D teams and projects by the network literature is puzzling. This book tries to rectify the deficiency.

A critical aspect of employing social networks to understanding actors' performance and outcomes is the recognition that they are multifaceted and operate on different levels depending on the type of relations that individuals maintain (Burt 1983; Hansen, Mors, and Løvås 2005; Tichy, Tushman, and Fombrun 1979). However, for the past three decades, the main focus of social network studies has been on the structure of the networks, to the neglect of the importance of the type of ties and their content (Adler and Kwon 2002; Monge and Contractor 2001). "Traditionally," Cross and Sproull observe, "network research has assumed that relationships can be appropriated for different purposes (e.g., friends can be sought for work-related information), and

so it is unnecessary to distinguish between kinds of ties or specify content in networks" (2004: 447). Similarly, the research on innovation has, generally, adopted this broad view and looks at the overall impact of social networks. It has been only very lately that this has begun to be seen as a potential impediment to furthering our understanding of the work dynamics in knowledge-based organizations.

Although not directly linked to the study of innovation, research conducted by Cummings (2004) and Cross and Sproull (2004) represent two empirical investigations in this direction. These studies identify several specific types of information in terms of content that people seek and share in work environments in order to accomplish their tasks. Moreover, Nebus (2006) argues that the network literature has exhibited a bias toward predicting outcomes by looking at a network's structural characteristics, while neglecting to attend to the question of how people form such networks in the first place. In particular, he contends that future investigations of knowledge-intensive environments ought to pay close attention to advice networks and calls for building a theory of how the latter are initially generated. At its core is the need to look at the processes through which individuals develop advice networks and the motivation behind their preferences for sources of work-related advice.

In conclusion, the review of the scholarship on innovation demonstrates that, in addition to paying attention to formal design characteristics, it is critical not to exclude from the analysis the impact that the informal structures exert on the process and the outcomes of technological innovation. Furthermore, the understanding of the role that social networks play will benefit not only from extending their investigation to the team and project levels, but also from studying both the structure and the content of various types of social relations. These include advice networks, the content of which reflects the work in an R&D environment. It is precisely how my study differs from the existing investigations that employ social networks to investigate technological innovation. Specifically, by examining the effect of two work-related advice networks, *technical* and *organizational*, which I constructed to denote the content of critical knowledge and information that is sought and exchanged in R&D organizations (Rizova 2002, 2006a), my book addresses Powell and Grodal's call for future research to "offer a more compelling analysis of the

specific ways in which networks shape innovative outputs" (2005: 79). To date, my research is the only empirical investigation to look at two complementary work-specific advice relations in the study of innovation and R&D projects.

RESEARCH ON PROJECT MANAGEMENT, TEAMS, AND SMALL GROUPS

The Literature on Project Management

A natural source of insights into R&D projects is the extensive body of literature that encompasses research on project management, teams, and small groups. Compared with the empirical investigation of technological innovation at the organizational and individual levels, R&D projects and teams have received somewhat less attention (Anderson and King 1993). The focus of studies on innovation at this level has been directed toward three major areas: the structural characteristics of teams, projects, and small groups; the climate conducive to group innovation; and the group processes themselves. Studies on successful innovation at the project level, in a manner similar to those at the organizational level, have also tended to produce conflicting results and to be inconclusive. One plausible explanation is that this could be a result of the lack of consensus on what "project success" actually means. As Griffin and Page argued, "[s]uccess is not just elusive; it is also multifaceted and difficult to measure" (1996: 478). Research has provided evidence that the definition and measurement of success are contextual (Balachandra and Friar 1997; Olson, Walker, and Ruekert 1995), and they could depend, among other factors, on the kind of strategy adopted (Griffin and Page 1996) and the type of innovation pursued (Green, Gavin, and Aiman-Smith 1995; Shenhar 2001).

What complicates the matter further is that neither success nor failure can be explained by a single factor, although scholarship has been seeking this elusive panacea for decades (Balachandra and Friar 1997). Indeed, Maidique and Zirger, who conducted the *Stanford Innovation Project,* argued that a range of factors, pertinent to both the firm and the project, tend to shape success, and therefore the quest for a magic bullet is not only unrealistic but illogical (1984). The project management literature of today faces a different dilemma—that of how to put some order into the vast number of factors that

have been suggested to explain the outcomes of technologically innovative projects. These include variables associated with the market, technology, the environment, the availability and utilization of financial resources, and the characteristics of the organization itself (Balachandra and Friar 1997; Brown and Eisenhardt 1995; Cooper 1979; Cooper and Kleinschmidt 1987; Griffin and Page 1996; Pinto and Slevin 1988; Rothwell et al. 1974; Shenhar 2001). A related problem is that studies have also tended to produce conflicting results. For instance, in a review of more than sixty articles on new product development (NPD) and R&D project success, Balachandra and Friar (1997, 1999) concluded that not only do the majority of the studies report results that do not build a uniform understanding of the fundamental forces behind success, but some of the findings conflict with one another.

Upon further investigation, the authors conducted a detailed analysis of nineteen empirical studies on R&D projects and NPD that reported no less than seventy-two success and failure factors. These they grouped into four main categories: market, technology, environment, and organization. Furthermore, half of these seventy-two factors were idiosyncratic to specific studies, and about three-quarters of the remaining half were only reported in one or two articles. As each article suggested between three and twelve factors to be most significant, Balachandra and Friar found few common elements in these studies. Ultimately they concluded that R&D and NPD success and failure are contextual, and they suggested that some consistency in the findings might be achieved by investigating projects against the background of three major axes: the nature of the innovation (radical or incremental), the nature of the market (existing or new), and the nature of the technology (familiar or experimental).

In a more recent review of the literature, Van der Panne, van Beers, and Kleinknecht (2003) examined forty-three articles published in peer-reviewed journals that report on the factors that have been found to explain the success and failure of innovative projects. The studies represent an amalgam of qualitative and quantitative research conducted between 1972 (when the notable SAPPHO project by Freeman and his colleagues took place [Freeman, Robertson, Achilladelis, and Jervis 1972]) and 1999. Van der Panne and his collaborators classified the factors into four major groups: those related to the firm,

to the project, to the product, and to the market. In nine of the forty-three papers they identified a large number of causes for success or failure. They also rank-ordered them. A closer analysis of the rankings in this subset led to the conclusion that there was a significant degree of similarity so far as the top ten factors were concerned, but very little agreement on the factors that ranked lower. Moreover, the remaining twenty-four studies reported findings that were either inconsistent or inconclusive. Some of the main explanatory variables in this category were support from top management; the type of organizational structure (functional, organic, matrix, or venture team); the degree to which a project is innovative; and the effect of the strength of competition.

They found a consensus of findings regarding the positive effect of the firm's culture, an organization's prior experience with bringing innovations to market, the diversity of the R&D team in terms of the balance between technical and marketing skills, management style, and the extent to which a project's demands for resources and the company's ability to match them were complementary. In sum, though, "[w]hile some studies claim a certain group of factors being crucial, other studies ignore the very same factors and claim very different factors to be decisive" (Van der Panne, van Beers, and Kleinknecht 2003: 310). As a result of such an abundance of explanatory factors, and the realization that the search for finding a single, all-important cause was naïve, the emphasis of empirical investigations in the project management literature in the late 1980s and early 1990s shifted to the discovery of sets of explanatory variables.

Despite systematically casting such a wide net to capture the factors that could explain success or failure, a puzzling but obvious omission from the project management literature on innovation in the past four decades has been the examination of the effect of social networks on group dynamics and outcomes. This lack of attention to social networks is even more surprising when considering that one of the most persuasive arguments about R&D project success has been based on the information-processing approach (Allen 1977, 1984). According to this line of reasoning, R&D "project effectiveness would be a function of matching communication patterns to the information processing demands of the project's work" (Tushman 1978: 640). An entire stream of research has been dedicated to studying both the sources

and modes of dissemination of internal and external communication in R&D organizations (Allen 1977; Ancona and Caldwell 1992; Katz and Allen 1985; Katz and Tushman 1981; Tushman 1977; Tushman and Scanlan 1981). The results from these studies established that "[t]he communication patterns on the high performing projects were systematically different than the communication patterns of the low performing projects" (Tushman 1978: 642). Ancona and Caldwell (1992) have shown that sharing knowledge outside of the group is positively related to performance. As the specialization of knowledge continues to grow, though, it is becoming increasingly clear that the transfer of knowledge, both within and outside groups, plays a fundamental role in an organization's ability to succeed at innovation (Argote, Ingram, Levine, and Moreland 2000; Argote, McEvily, and Reagans 2003). To this end, including social networks in the study of the sets of factors conducive to success on technical projects into the literature on project management carries great potential for generating powerful insights into the mechanisms that groups develop for sharing and transferring knowledge.

Teams and Small-Group Studies

Yet another branch of scholarship that has been very influential on the study of group dynamics is that focused on teams and small groups.[2] Over the past few decades, it has produced compelling evidence concerning the predictive value of individual, environmental, and group processes to team and group effectiveness. Historically, the main variables of interest to the researchers from this tradition have been size, leadership, group cohesiveness, goals, and motivation (Guzzo and Dickson 1996). Currently at its central focus are the issues surrounding team composition, and especially the role of diversity in it.

A large number of studies from this tradition have concentrated on the features of team design. Stewart (2006) conducted a comprehensive meta-analysis of ninety-three articles published in peer-review journals that looked at the relationship between aspects of design and a team's performance. His analysis covered research published over thirty years, up to 2003, and included both quantitative and qualitative studies. The great majority of these authors approached their investigations from the widely adopted "input-process-output" framework for teams (McGrath 1984). Such studies focused on four broad de-

sign categories: group composition (members' characteristics, heterogeneity); task design (mechanisms for differentiation and integration); the meaningfulness of tasks and the degree of team authority; and the organizational context (leadership and perceptions of leadership support). Stewart reached the conclusion that the correlation of these design categories to team performance was different for production, project, and management teams.

In a study of 626 individuals in forty-five production teams, Stewart and Barrick (2000) found a strong relationship between team structure and team performance. More specifically, on the basis of prior research as well as their own findings, the two investigators concluded that "structural characteristics related to the allocation of tasks, responsibilities, and authority do indeed influence team performance" (2000: 144). Interestingly enough, though, at the center of this literature is a focus on the formal structural characteristics, with the gross omission of the informal structures. Even when team processes are discussed, the analysis fails to incorporate the specific ways in which human agents get involved in social interactions. In fact, one cannot help noticing that in the same manner in which the scholarship on social networks has neglected teams, social networks have been neglected by the literature on teams and small groups. Furthermore, in a review of the literature on teams in organizations, Ilgen and his colleagues (2005) have also observed that it has been only recently that a few studies have employed social networks to better understand the patterns of interactions between, and within, teams.

For instance, the study of innovative teams has reached a consensus that in addition to the importance of having vision, clear task specifications, and support for innovation from other members of the group, the fourth critical group of variables has a lot to do with what West (1990) terms "participative safety." He argues that "the more people participate in decision making through having influence, interaction, and sharing information, the more likely they are to invest in the outcomes of those decisions and to offer ideas for new and improved ways of working" (Burningham and West 1995: 107). Given that a high level of participation in small groups has been strongly linked to trust, and that the latter was shown to emanate from social networks, the failure of this literature to incorporate the examination of the informal relations is indeed difficult to understand. Consequently, one of the contributions that my book

offers to the literature on teams and small groups is that it discusses the *specific* ways in which different social relations affect team performance, while integrating both formal and social network structural properties into the analysis of the allocation of tasks and authority.

Taken as a whole, the focus of the team and small-group literature has been on five groups of factors: environmental factors, processes and incentive systems, the characteristics of the team members, the identification of the innovators, and the composition of the teams in terms of heterogeneity, as well as the type of organizational culture that would enhance innovation. The assumption in this literature seems to be that once the right team and individual attributes from these groups are identified and put together, successful outcomes are more likely to follow. Accordingly, as far as team composition is concerned, the emphasis has been on mixing and matching attributes. However, the specific relations that individuals enter into and develop in small groups, during the process of knowledge creation and sharing, is absent from this tradition. My study differs from this approach by specifically investigating the different types of social relations people develop during the process of innovation and the ways in which these relations contribute to the shaping of the outcomes of projects and teams.

Furthermore, three distinct and exciting recent developments have taken place in this fertile body of scholarship that, without a doubt, are going to inform and shape profoundly the direction of the future research. One such development has been the recognition that small groups and teams have been generally investigated in a static manner divorced from their immediate context (McGrath, Arrow, and Berdahl 2000; West, Hirst, Richter, and Shipton 2004). The overwhelming majority of the existing models that have been built to explain team performance, as I suggested earlier, are based on McGrath's (1984) "input-process-output perspective" (Stewart and Barrick 2000; Ilgen, Hollenbeck, Johnson, and Jundt 2005). The model rests on the assumption that structural characteristics (inputs) exert an effect on the team's processing abilities and, consequently, shape the outputs. Inputs constitute the "knowledge, skills and abilities of group members; the composition of the team; and aspects of the organizational context such as the task and the associated objectives, reward systems, information systems, and training resources. Process

refers to the interactions among group members, information exchange, and patterns of participation in decision-making, leadership, social support, and sanctions for group related behaviour. Outputs include the products of the group's performance. . . ." (West, Hirst, Richter, and Shipton 2004: 273). An impressive amount of research has been conducted within this tradition, and extensive knowledge has been generated about groups, including several classic studies, such as those undertaken by McGrath (1984) and Hackman (1987). More recently, though, the model has been seen as an inadequate framework from which to understand the changing nature of teams, their fluid boundaries, and the intensity of interaction within them. A new way of thinking has emerged that calls for a reorientation toward the study of small groups and projects. The need for such a shift is based on the realization that the "input-process-output" model is a static one and implies a linear progression from inputs to process and ultimately to outputs (McGrath, Arrow, and Berdahl 2000; West, Hirst, Richter, and Shipton 2004; Ilgen, Hollenbeck, Johnson, and Jundt 2005). Indeed, one of the major criticisms against using it to understand and explain team behavior has been precisely that it does not incorporate a feedback mechanism (West, Hirst, Richter, and Shipton 2004).

A second distinct change in the focus and framework in this tradition involves a preference for multilevel theoretical and empirical research. The obsession with the outcomes of team performance, so characteristic of this stream of research up to the late 1990s, has now shifted to a heightened attention toward processes that mediate the relationship between inputs and performance. Moreover, the emphasis has changed to answering the question of *why* they have this effect. Yet a third clearly marked trend that has developed over the past ten years is expressed in the appeal to consider groups as complex, adaptive, and dynamic systems. It promotes the empirical investigation of groups and group processes as they develop out of the interaction with other levels of analysis and within a particular context (McGrath, Arrow, and Berdahl 2000: 95).

It is important to acknowledge, therefore, that groups and projects are not a mechanical collection of individuals with the appropriate technical skills, as viewed by many organizational experts, but are actually much more than this. They are entities in their own right, and they possess unique dynamics. Otherwise all groups and projects who share structural properties, and are in

possession of the relevant human capital, would be more or less equal performers. Thus, examining the effects of both the formal and the social network structures on the creation of these dynamics seems to be a good place to start. As I have shown earlier, the social network perspective has been applied to the study of innovation at the inter- and intraorganizational levels, but hardly at all to study projects. It is unrealistic, however, to expect that the interactions at these higher levels and the dynamics that they generate will be mirrored in small groups and projects. The intensity of the social interaction is much higher in them, and so it is plausible to expect that social networks and their properties will create somewhat different effects on group dynamics and, consequently, on performance. These relationships, in turn, are also affected by dynamics at the level of the organization that are created as a result of their peculiar structural properties, both of a formal and informal nature, and informed by systems of shared meanings.

In conclusion, the examination of the current scholarship on project management, teams, and small groups has raised several important observations. Our understanding of how to account for success and failure on technologically innovative projects will greatly benefit from approaching projects as small, complex systems with dynamics of their own. These dynamics are to a large degree context-specific and process-based, and not only result from interactions prescribed through formal channels but are also motivated by the specific patterns of social relations that team members establish. To this end, an investigation of the sets of factors associated with the processing of knowledge within innovative organizations is paramount. So, too, is achieving a better understanding of the dynamics conducive to the success of R&D projects by simultaneously exploring positive outcomes from both vantage points—the formal and the social networks' structures. Such research should incorporate multiple levels and be sensitive to the specific context within which the work is undertaken.

My own research strategy shares this holistic approach. Although structural factors at the project level are my main focus, I also pay close attention to organizational processes and cultural variables, which in combination with the structure played a critical role in shaping the outcomes of six technical projects in an R&D organization.

FORMAL AND INFORMAL STRUCTURAL EFFECTS
AT THE TEAM AND PROJECT LEVELS

As I have shown earlier, scholarship stemming from the fields of the sociology of organizations and management science has produced an impressive body of empirical evidence concerning the ways in which formal, as well as informal, structures affect organizational performance. Results from earlier small-group studies, too, have demonstrated how the structure of the formal (Guzzo and Shea 1992) and informal networks (Shaw 1964) shapes group outputs. Despite the acknowledgment that "neither perspective by itself yields satisfactory understanding of social and [organizational] phenomena," it is not unusual for the two structures to be viewed as "antithetical and even irreconcilable" (Knoke and Kuklinski 1982: 10). It has also been recognized that, by and large, managers are unaware of the extent to which the "invisible" informal relations underpin the formal activities (Cross, Borgatti, and Parker 2002).

A number of scholars have documented the interpenetrating relationship between the two structures (Blau and Scott 1962; Dalton 1959; Nee 1998). Mintzberg, for instance, has acknowledged that "[t]he formal shapes the informal, while the informal greatly influences what works in the formal, and sometimes even reflects its shape to come" (1979: 53). Although quite limited, some research effort has been directed at the investigation of the ways in which the formal and informal structures relate. The focus of this work has been mainly on how the formal organization influences the development of the informal relations (Blau and Alba 1982; Brass 1984; Krackhardt 1990; Shrader, Lincoln, and Hoffman 1989); how the informal structural sources of power supplant and help overcome the inefficiencies of the formal arrangements (Stevenson and Gilly 1991); how the combination of personal sources of power (for example, one's level of expertise as a result from experience, seniority, and education) and structural centrality affects the involvement in innovation (Ibarra 1993); and how centrality in an organization's formal and communication network structures affects the perception and use of power (Brass and Burkhardt 1993).

For instance, while examining the information flow in a hospital, Stevenson and Gilly (1991: 918) looked at whether a "preexisting network of ties between individuals inhibit or facilitate the passing of problems to formal problem

solvers." They found that hospital managers preferred to use personal connections to relay information about patient complaints to problem solvers. Ayers, Gordon, and Schoenbachler (2001) looked at the role of formal and informal controls on the effective collaboration and integration of key functional areas (such as R&D and marketing) and, thus, on the successful development of new products in a study of 152 managers. Their results reveal a positive relationship between the collaborative interaction and the success of NPD. Such cooperative relations, they note, are fostered by decentralizing the decision making and clarifying the roles in the process of developing new products.

Although the investigation of the interplay between structural sources of power has not been the explicit aim of these studies, they produced powerful insights at the individual level. The interplay between the formal and informal structures through status positions and roles, however, and the ways it affects performance and outcomes, particularly at the project level in knowledge-based organizations, is not well understood (Smith-Doerr and Powell 2005). There has yet to be developed and tested a different approach to organization and project design: a framework that incorporates the simultaneous effect of both structures in a dynamic fashion, or, to borrow a phrase, "how they structure one another" (Thomas 1994: 5).[3] It is also necessary to link their effect to action and the production and reproduction of systems of meaning, power, and cultural norms (Giddens 1979; Vallas 2006). Nohria and Gulati (1994: 529) have called for such a synthesis. "What we need," they assert, is to "focus on how the formal and informal structures are interrelated and influence each other." Recently Jonathon Cummings, too, has pointed out that theory and research on the types of diversity that "influence the value of knowledge sharing in work groups" need to pay more attention "to member differences in organizational affiliations, roles, or positions" (2004: 352–3). What is more, no existing studies on R&D projects and their outcomes integrate the combined effects of both structures through an investigation of positions, status, and roles.

Given the growing attention in the literature to managing the informal relations in addition to the formal ones in knowledge-intensive organizations (Cross, Borgatti, and Parker 2002), there is a need to fill this gap. By examining in a systematic fashion the assignment practices and processes at an R&D laboratory, and analyzing how roles and positions in the formal and in two

complementary work-related advice networks interrelate, the research described in this book addresses two main concerns. First, whether it is possible for technical organizations to develop a system that will enable the continuous and fluid sharing of knowledge and information across a broad spectrum of stakeholders and, second, whether human and social capital could be formally aligned with the organization's goals and challenges.

My findings are derived from the analyses of qualitative and quantitative original data collected on six projects at a *Fortune* 500 company dedicated to technological innovation. I conducted forty-three one-on-one, open-ended and semi-structured interviews with the laboratory director, the managers of the six projects, and the vast majority of the members who worked on these projects. In addition, I also collected quantitative and social network data on four different types of social relations that people at the lab are likely to enter into during their daily work. I arrived at the conclusions by employing three different methodologies to analyze the data: an inductive approach to process the interview material; Ragin's qualitative comparative methodology (2000), which combines qualitative and quantitative data to identify the set of critical project success factors as well as to test interactive relationships; and social network analyses.

THE ORGANIZATION OF THE BOOK

This chapter has raised the basic questions addressed in the book. I started by briefly outlining the importance of technological innovation for society. It has been followed by a discussion of innovation as a process in the light of the demands and constraints it imposes on management, particularly for knowledge-intensive organizations operating within the global economy. Then I discussed the three bodies of scholarship from which the main questions informing this research developed.

Chapter 2 begins by providing background information on the setting for my study—the company and the R&D facility. It then presents and discusses the results from the inductive analysis of the qualitative data, stressing the four factors found to be critical to the success of the technical projects under investigation in the laboratory. In Chapter 3, I employ the results from the qualitative and the social network analyses to demonstrate the specific

ways in which the presence or absence of each of the four factors contributes to the degrees of success of the six projects. The research uncovers the unique method of allocating tasks, responsibilities, and authority on projects at Global East and discusses the cultural norms and work processes that support and reinforce it.

Next, in Chapter 4, I elaborate on the central themes by reconstructing the histories of three of the study's projects—*Delta, Alpha,* and *Beta.* I discuss these projects' formal and advice network structures, as well as the social dynamics they create in the light of the findings reported in the previous chapters. An analysis of *Delta* is particularly interesting, as the project started out as a failure but was turned into a success as a result of recombining the four conditions. Then the story of the highly successful project *Beta* is contrasted with that of *Alpha,* a low-success project, to demonstrate how their different structural arrangements produced very different outcomes.

In the final chapter, I discuss the contribution that this research makes to our understanding of innovation, to the process of knowledge sharing and individual and organizational learning, and to the role of social networks in it. I also consider the implications of the study for the managers of R&D units and technical projects and teams. This concluding chapter emphasizes how the mechanisms revealed in my research can be applied to other organizational settings, not only to achieve an understanding of their social dynamics but also to demonstrate how best to mobilize their human and social capital to foster the creation of knowledge and innovation.[4]

2

SUCCESS IN TECHNICALLY

INNOVATIVE PROJECTS:

FOUR CRITICAL FACTORS

"The traditional sequential . . . approach to product development . . .
may conflict with the goals of maximum speed and flexibility.
Instead, a holistic or 'rugby' approach—where a team tries to go the
distance as a unit, passing the ball back and forth—may better serve
today's competitive requirements. . . . Under the rugby approach, the
product development process emerges from the constant interaction
of a hand-picked, multidisciplinary team whose members work
together from start to finish."

Takeuchi and Nonaka 1986: 137–38

THE SETTING

The site of this study is the R&D unit of a U.S.-based *Fortune* 500 company,
which I refer to as "Global East."[1] Global East ranks among the top 200 Global
Fortune companies and has been a leader in its core business categories for the
past fifty years. It employs over a hundred thousand individuals in facilities in
nearly thirty countries. The company's commitment to research and development is reflected in the sheer size of its R&D operation—about two dozen laboratories that employ more than two thousand researchers. Only in the past
decade were two new R&D facilities created; this study is conducted in one of
them. The R&D laboratory was established to respond to the growing needs of
the company's largest and most profitable business division. For the period of
2000–2002, at the time of the study, the business division marked record sales
that accounted for 6 percent of the company's total sales. The two major foci
of the R&D laboratory are to respond to a need for research, development,
or testing from its respective business division or another R&D laboratory
within the company and to originate, develop, and test new product, process,
and evaluation ideas. The implication of the fact that the R&D facility was

created to respond to the research, development, and technical needs of one of the company's divisions is that about 50 percent of the lab's projects are of a routine nature involving technical services. The duration of such projects is from two months to less than a year, and they are normally staffed with one, two, or three lab members. On the other hand, the remaining portion of the company's portfolio consists of projects that aim at achieving "a greater degree of innovation." These last between one and five years and could comprise from five to over fifteen project members, some of whom may also come from other divisions within the company.

The facility[2] employs about fifty scientists, engineers, technicians, management personnel, and trainees, of whom 10 percent are women. The average age in the laboratory is forty-three years; age ranges from twenty-seven to sixty-five. About 60 percent of the personnel hold advanced technical degrees. The laboratory is headed by a director and consists of three major groups that are run by group managers. The turnover rate is low; about 80 percent of all people employed at the lab at the time of the study had been there since its establishment. In the three years preceding my investigation, one senior engineer, one junior scientist, and a trainee were hired; a junior scientist transferred from an overseas R&D lab; and one technician transferred from another division at Global East.

Through a combination of qualitative, quantitative, and social network analyses I studied six technologically innovative projects at the facility.[3] The laboratory director categorized the outcomes of the projects based on whether or not, in his opinion, expectations had been met on (1) potential financial returns, (2) satisfying the technical parameters, and (3) staying within budget constraints (Keller 1994, 2001).[4] As none of the projects investigated in this study were terminated prior to its completion, the director argued strongly that their outcomes would be more accurately depicted as a continuum, rather than as a "success" and "failure" dichotomy.[5] Hence, he labeled the outcomes of the projects that exceeded the initial expectations as "high success" and of those that just about met the initial expectations as "low success."[6] Summaries of the six projects, which I refer to as *Alpha, Beta, Gamma, Delta, Epsilon,* and *Zeta,* are provided in Appendix 1.

The argument I advance in the book is based on the analysis of qualitative,

group manager depicts the association between open communication and work efficiency clearly:

> We have to know what we do and how. . . . When something comes up and maybe it is not her [refers to a researcher] project area—she helps. And it is pretty much like this with everybody here. Nobody is isolated. We simply can't afford it because there is so much to be done and unless we pool our brains, it just doesn't happen fast enough. To be able to give sound advice, though, each of us has to be aware of who works on what problem, what went wrong, what went right, who else might have worked on a related problem and so on. So, we talk about work . . . a lot.

In a business environment time is money; it is the nature of the beast. Therefore, it is not simply desirable but essential for "things to happen fast"; otherwise the opportunities disappear. Constant and unrestricted exchange of work-related knowledge and information is one way to secure this.

The researchers, engineers, and technicians highlight the same benefits when explaining the significance of open communication. A researcher on project *Beta* captures the scientists' views on the matter and, in addition, connects open communication to the issues of respect, work organization, and the achievement of the project's goal. His comments are worth quoting at length:

> It makes a huge difference when you know that anytime you need to discuss something with your manager, she will listen. I feel free to talk to my manager about anything anytime. If she does not have the time, she will tell me, and when she is free, she will come to me. . . . You asked me about reporting formally. . . . We get together to discuss things, but I wouldn't say that we "report." We discuss the problems and the agenda for the day and for the week, but reporting formally . . . No . . . You see the difference . . . ? People who make appointments here are people from the outside. The nature of the work here will not tolerate it. You see, . . . you have to make decisions quickly in this environment. We are not just research, development, and testing. . . . We are a business [organization]. You can't wait for your "appointment" to talk to a manager or to a colleague. Things change quickly here. If you are scheduled to use a machine and you come to a problem, you have to solve it right away because you have a limited time on the machine. Someone else is scheduled to use it tomorrow. . . . You understand . . . ?

At the same time, many of my interviewees pointed out that a shift in the communication patterns takes place as a project moves through its various

stages of development. The shift concerns the frequency and the means of the information exchange on the progress of the project. Frequent dialogue takes place during the stage when the technical goals are not yet clearly defined, when a number of approaches to solving the problem are tried and there is a lot of ambiguity as to how to proceed. Once answers to these questions are found and the technical specifications are defined, everyone knows what his or her role is. The manager and the project members meet less frequently at that time and on a need-to-know basis. The project members are provided the autonomy to carry out their tasks, and they report in writing on the results from the experiments and tests that they carry out as soon as they have them. As the director explains, "[T]he challenge of every manager in an innovation group is to create each project in its initiation phase and then to attach specificity to it. . . . To facilitate each project to kind of be moved into this direction . . . Because if you make every project uncertain the people will go crazy." At this stage of a project, "formal communication" and "reporting" means that there is a routine in place for recording the results so that the managers can monitor the progress and make changes if necessary. To my respondents, this did not mean "scheduled meetings," "presentations," or "bags of paperwork." The open communication patterns and the solicitation of help and information to resolve an issue at the laboratory level, however, do not change and are not specific to a particular stage.

A researcher who worked on *Epsilon* elaborates on the importance of open and direct communication in terms of building trust and showing respect to colleagues. Again, his statement goes beyond pointing out the social value of what is often referred to, in managerial parlance, as "soft stuff." He talks about how open communication channels, devoid of bureaucratic barriers, create an organizational culture that deals efficiently and promptly with technical issues and questions.

> All of us work on multiple projects. Sometimes our heads spin. . . . Oftentimes [laughs] . . . None of us knows everything. Sometimes you just need to ask someone who had worked on a similar problem what to do in a situation . . . To . . . sort of . . . help you a bit . . . It could be a technician, an engineer, or the manager. . . . Doesn't matter. And you need to solve this problem *now*. They help you on this one, you help them on something else. If you ask a person and he takes a long time to come back to you or say . . . , he is kind of grumpy about it,

next time when he needs your help you will do the same. . . . You can't say, "It's not my job, I don't report to you." . . . This is a small facility. . . . Attitudes like this will create a very bad atmosphere. . . . Who needs this? . . . We just need to be aware twenty-four-seven of what's going on in the lab because there is constantly something to fix, a question to answer . . . so, you have got to know where to look for the answer.

Olga, a group manager and one of *Beta*'s project managers, echoes this sentiment while adding that this attitude is not bound to a specific project. Rather, it is characteristic of the way in which the laboratory functions. She makes it clear that "proprietary attitudes" with regard to "working on *my* project" and "caring about the success of *my* project only" are discouraged in the facility. "Just *doing* my job" or what is even worse, "doing *just* my job" is the type of attitude that is likely to guarantee that one *does not* move up in the formal and social hierarchy in the laboratory. Simply because a person is not formally assigned to a project does *not* make it acceptable to refuse helping a colleague if he or she has the expertise. At the same time, the question of whether one has the time to spare on figuring out somebody else's problem is not even brought up in the discussion; it is taken for granted that no one does. It is expected, however, that time will be made. Olga describes this relationship in the following way:

> Everybody is so overloaded with projects that they appreciate any help. And everybody knows when someone comes to you and says, "I am having some problems. Could you help me?" people are more than willing to help because they know that they will also be in that spot. It is the way the body fights a disease—by mobilizing all of its forces and concentrating all its efforts there.

Thus, open and direct communication that cuts across organizational and project boundaries is assigned in this laboratory an even greater meaning and significance—that of the organization and project's survival.

I probed whether the respondents and I share the same understanding of the meaning of "open and informal communication on the project" as opposed to "reporting in a formal way" by asking them to offer concrete examples. This is what one of the researchers stated:

> She [the project manager] did not ask me to [report], but I wanted to make sure that she was okay with it, so I would tell her what I was doing every day. We had

offices next to each other and we would chat. . . . So, we discuss. It is like an open discussion. Just everyday discussion . . . what needs to be done . . . No, there is nothing formal about it. Just yell across the wall talking to her. It's like an open cubicle type, so she was in a cubicle and I was in the next cubicle. . . . I would tell her "this is what I am planning on," and if she had some ideas, she would throw them back at me. . . . So, it was an open discussion.

A technician on *Gamma* eloquently depicted the complex web of issues connected to the informal and nonhierarchical two-way communication patterns on projects as it relates to the manager's effectiveness in achieving the organizational goals. Among those are the themes of building mutual respect and an appreciation of the members' contributions across the hierarchical lines of authority and learning from one another. The technician stated,

Before I started here [working for the laboratory] I was with another division [in the same company]. The engineer there was an engineer from hell. He never asked for our input. He comes and tells you how to do your job. It was very aggravating because often we knew that what he was asking us to do won't work, but we had to do it because he was the boss. . . . It did not work out at all. . . . It created a lot of friction. . . . You see . . . he went to college for a number of years to learn about these things. He has a degree. . . . But while he was there reading about these things, I was here doing these things [technical procedures and tests]. . . . In this business experience counts. . . . a lot. Fred is a different kind of person. Everybody wants to work with him—trainees, other engineers, technicians . . . everybody . . . It is really very easy to work with him because he is very competent about machines, technically I mean, and about people. He just treats you right. You are not afraid to tell him that you don't know something or to ask a question. . . . Not afraid to tell him that there has been a problem. He trusts us and asks for our input.

Thus, direct and informal communication that cuts through all formal hierarchical lines acts as an invisible rapid response mechanism for addressing the ever-present problems in a technical environment. It also serves as an awareness mechanism. Managers, engineers, technicians, and researchers keep each other informed about the problems they stumble upon and the solutions they find by putting their trust in this mechanism, rather than fearing punishment or "looking stupid" or "incompetent." This avoids duplication of work not only on a project but at the facility and helps solve problems quickly (Bohnet 1997). Further, turning this communication style into an informal social norm within

the laboratory, and on each project, as Olga's account confirms, serves the purpose of releasing tension. In a busy and competitive industrial environment, it is reassuring to know that, while everyone is responsible for his or her own work, when advice or help is needed it will be provided; that there will always be someone who knows where to look for a solution, and whom to ask. This free exchange of knowledge and information creates a perpetual learning environment, as people who hold different positions within the organization and on the projects are continually in contact with one another and aware of each other's work. They constantly learn from one another. That gives them a sense of importance, and they are willing and motivated to relentlessly expand their expertise so that they, in turn, can offer help on even more complicated technical matters and get the social prize: to be respected as competent team players.

Critical Success Factor 2: Technical "Stars" Have Been Assigned to the Project

Good researchers do not necessarily make good managers and vice versa. Oftentimes, brilliant scientists and engineers are promoted to managerial positions and they perform abysmally at this level. As Putts' law encapsulates it, "Technology is dominated by two types of people: those who understand what they do not manage, and those who manage what they do not understand." Such organizational truisms are fittingly depicted by the well-known book *The Peter Principle* (Peter and Hull 1969), according to which people in hierarchical organizations tend to get promoted to their level of *incompetence*. Yet another truism is that an organization whose sole purpose is to succeed at technological innovation would not achieve its goal in the absence of either technical or managerial talent. Hence, it was not surprising that the critical importance of the skills, abilities, and expertise of both to a project's success were unmistakably recognized by all members of Global East's R&D laboratory. There was not a single respondent who did not acknowledge the vital nature of both technical and managerial competencies for the achievement of a successful project outcome. Regardless of their position within the organization and on the projects' formal structure, none of the interviewees had difficulty differentiating between the two types of competencies or understanding their instrumental role

to the success of the projects. Furthermore, none of them had difficulty identifying the project members who best fit the categories of technical and organizational "stars" or providing examples to substantiate their answers as to why they regarded these individuals as "instrumental" to the success of their respective projects.

Among the most frequently occurring descriptive terms used by Global East's R&D personnel to define the meaning of "technical" competencies and skills are "technical experience," "technical expertise," being able to make "the most unusual connections," being a "technical genius" or a "technological wizard," knowing "what might work and what is worth trying," knowing "this machine like your five fingers," and being "intuitive, but in a mathematically precise way." Not surprisingly, in all narratives the technical skills and competencies are directly connected to the projects' success, regardless of the positions the respondents hold in the formal organizational structure.

The direct and indirect ways in which technical skills and competencies affect the success of projects are reflected in three reoccurring themes that emerged from the interviews. One theme draws a link between the quality of the technical talent on a project, securing good communication flow and exchange of information, and ultimately achieving greater efficiency. This, according to the projects' members, is accomplished when people with high technical skills and expertise are able to speak and translate from different "technical languages." In the following narrative, a technician describes what makes Fred one of the "technical stars" in the lab. He stresses the fact that Fred's good grasp of the technicians' work, in addition to his own technical expertise, makes the communication not just easy but clear. Thus, the possibility for misunderstandings is reduced and, as a result, duplications of tests and procedures are avoided and the collegial atmosphere is maintained. Here is his account:

> Fred has an excellent background about machines, technical and mechanical. He has a lot of experience, some fifteen years of design work. That is why when he talks to us [the technicians] we know what he means, and he understands us. Sometimes you have an engineer who is great in designing but knows very little about our neck of the woods. But with him [Fred], we don't spend our time on constantly clarifying the *whats* and the *hows* . . . . [Laughs . . .] It also means we don't waste time filling the CYA [cover your a—] file cabinet. . . .

Second, the value of technical competence on projects is linked to the excellent reputation that the "technical wizards" have and the positive consequences it generates, not just for a particular project but for the entire organization. The lab director links the technical reputation of the laboratory's "technical stars" to the success in projects and to the laboratory's achievement:

> People like Fred are respected quite a bit. These people [points to business units] come to him, and that makes this place more successful. This way they utilize us. . . . Success here depends on how much the engineer or the chief scientist makes himself indispensable to the needs of this group [the business unit] . . . his expert power. So his success and the success of the organization are based on expert power. And then, people talk about it. They know we have the expertise.

The value of the "technical stars" in an R&D organization clearly transcends time and a project's boundaries. It is understood to be not only vital for a particular project's success, but through it—as a mechanism that contributes to the present and future technical achievements of the R&D facility—a factor that enhances the laboratory's reputation within the organization and, by extension, the reputation of the company within the industry. In addition, it is a mechanism of earning the respect of the corporate management and "a sort of guarantee" for more challenging and visible projects for the facility in the future.

The third recurring theme connects the importance of having people with exceptional technical talents and skills on a project to the creation of a much-needed sense of stability in a fast-paced, highly competitive industrial environment. In critical situations, which is a fact of everyday life in industrial R&D labs, the people with exceptional technical skills and expertise are looked upon as time and effort savers, as "efficiency manager[s]," and as "islands of sanity." These themes are exemplified in the narrative offered by a technician who worked on project *Beta*.

> Sometimes it is just a matter of trial and error. You go downstairs and run hundreds, . . . thousands of tests. If the engineer is not very competent, he will ask you to do things that will turn into nothing. Sometimes, you just don't know. . . . But a good engineer has a clue, . . . an intuition about what could . . . , what just might work. . . . Otherwise, we [the people on the project] will go bananas . . . especially when you are working in a new territory kind of thing. They both [the two project managers] were great about it. They know the work and won't just

try anything. . . . She [Olga] was concerned about deadlines. . . . If it wasn't for the years of experience, we would be still testing. . . .

Here the respondent attributes *Beta*'s spectacular success to a large degree to the technical experience and expertise of the two project managers—Olga and Tonya. Given the history of *Beta*, this statement makes a lot of sense. *Beta* was a very high profile project on which not one, but two vice presidents were pushing for results, and not just daily, but hourly. Financial resources were abundant for this project, and the corporate management made it explicitly clear that these could be easily negotiated and readjusted further if and when needed. The timetable, however, was not. The laboratory and the company had a lot at stake by undertaking the project, and as Olga put it, "they could not afford to fail." In a situation such as this, waste of time cannot be justified even if the additional time for testing and exploration results in the creation of new knowledge. The pressure is colossal, emotions are taxed, and there is a great need to trust one's technical experience, while not blindly relying on it. This need is met at Global East's lab by securing open communication channels and assigning people to projects with no-less-than-excellent technical, organizational, and social talents, skills and capabilities; people who have proved themselves in earlier battles; people who are respected and trusted. The utilization of these mechanisms creates a work atmosphere that conveys to the project members, to the laboratory, and to the company's management the message that although hectic, both the daily work and the project's progress are under control. This brings me to the third factor that the analysis of the interview data suggests makes a vital contribution to the successful outcomes of innovative projects—having people with managerial "know-how" and skills on the project.

Critical Success Factor 3: Managerial "Stars" Have Been Assigned to the Project

As was the case with the technical skills, the utmost importance of the organizational, or managerial, expertise for a project's success was equally highly regarded and unequivocally acknowledged by all interviewees, irrespective of the structural positions they hold in the laboratory and on their projects. Here, the word *managerial* does not necessarily refer to an individual's ability

to lead and oversee a group. Rather, it refers to certain competencies. Furthermore, none of the lab members had difficulty defining the meaning and significance of this factor to the success of a technical project or identifying the people who possess such competencies—"the managerial stars." The contribution of having an individual with organizational capabilities and skills on a project is explained through the themes of efficiency, smooth project functioning, and proper deployment of human capital and social capital. An intriguing observation with regard to the discussion of this set of skills is that about half of the respondents distinguish between managerial skills and competencies relevant to the success of the R&D organization and those relevant to the success of a project. For instance, in the following quote, an engineer defines the organizational competence at the project level as

> [The] ability to quickly find the right parts and materials, to get the best provider, to make sure the equipment is available, to know everything that there is to know about a particular product line, to procure the things we need, to organize the process so that the work is uninterrupted.

An account by a group manager shows that across hierarchical lines people at the laboratory share a consistent and clear understanding of how such competencies come to be used by a project manager. Here, for instance, is how Olga sees her role as *Beta*'s project manager:

> So and so needs help with [names a technical procedure]. So I might say, I know that this person [points to a lab member] will need more training in [names a technical area] so I'm going to ask them to go and work with Patrick and Tonya. They will sit down as a group, work out what needs to be done, and the person who needs the training will do it. So this is simply my job. . . . To take the resources and start playing a shell game and making sure that this project is a five-week project, a two-week project, a five-month project, a year project. . . .

Olga's account is very similar to the one offered earlier by an engineer. She stresses all the indispensable elements of her daily work as a manager—knowing what needs to be done, knowing who can do it, knowing how to close any gaps, keeping the project on focus and running, meeting the deadlines, and making sure that the technical expertise in the laboratory is continuously reproduced and further developed in the process.

At the level of a project, Nick is clearly the "organizational star" for he "keeps the place running" and "he does it in a way that is fair to everyone." He is able to accomplish this "because he has the right connections in the company and outside [with manufactures and suppliers]," "people would talk to him about anything," and "he knows this organization upside down." Furthermore, he has the right attitude because "he is here to solve problems, not to make you fill forms." A technician describes in the following way the importance of having good organizational skills in general, and Nick's in particular, to the efficiency with which the whole laboratory functions and, by extension, the projects he manages:

> He contributes in a very big way because he coordinates everything and makes things run smoothly. When you work on several projects at the same time and need different equipment and suppliers and skills and so on, you could not be more thankful that Nick is in charge. He knows how to schedule you for testing—how much time you need, what could go wrong . . . that kind of thing, how to come up with some predictability in this business. . . . Keeps you on track. . . . He saves a lot of time.

At the company level, on the other hand, the laboratory members defined the set of organizational competencies and skills as "being well connected within [Global East] and outside," "knowing how to play politics," "upholding the image of the facility," and "knowing whom to talk to." They identified Tom and Olga as the most prominent laboratory members who possess these skills at the level of the organization. The impact of their skills was traced directly not only to the success of the R&D facility but to the projects as well. For instance, the following statement offered by a researcher epitomizes the understanding of the laboratory's members of the indispensability of the organizational expertise and the contribution it makes to a successful project outcome:

> Tom knows what projects will make technical sense and how to navigate them politically. He is very connected. . . . When he speaks, people listen to him. . . . *Zeta*, for instance, is entirely his idea. . . . This project has been very good for us. It gave us a good name across the entire company, here [in the United States] and overseas.

Thus, it is not sufficient that a particular project is properly managed. Rather, the seeds for success are planted much earlier; it is just as important

to select and attract projects that make strategic sense for the laboratory and carry the potential to strengthen its reputation. To accomplish this goal, one must possess both the competencies and the "clout" needed to convince the corporate decision makers of these benefits as well as the technical and marketing vision to recognize a project's vitality.

Most scientists, engineers, and technicians in the laboratory regard the technical and organizational skills as interrelated; as reinforcing each other. As one of the managers explains why his project was rated as "highly successful," he stresses the connection between the technical and managerial abilities: "In the final analysis, what made this project so successful was the excellent planning and coordination and the skilled people. If we needed help we hired a consultant." Thus, he succinctly links the undivided contribution of technical and organizational competencies to the project's outcome. Nick, a "managerial star," who often co-leads projects with Fred, a "technical star," understands it in the same way himself. In describing how the two of them worked on *Gamma,* he makes the point:

> We understand one another and we save a lot of time going back and forth. Fred tells me what the technical issue is, and I find the person with the right skills and I make sure this person is not going crazy working on ten other projects. In other words, I make sure that Fred got this person's mind and his hands 100 percent. . . . I know which manufacturers to contact, what it takes, and how to do it. I make sure we don't fall behind and that we stay within budget. Fred can concentrate on the technical side.

Fred's account of how he and Nick co-managed this project is rather similar. Yet he places the emphasis on the constant consultation and mutual decision making that took place between the two, which were aided by the fact that they know enough about each other's areas of expertise.

> Nick and I started to look at suppliers of this equipment—where can we get these machines—and we started working with purchasing, identifying some machine manufacturers. And then, let's see . . . , at the same time after working with Leny and working with our field engineers and going on field trips, I determined what it was we needed specifically . . . for the machine's capacity. So then I wrote up a specification for the machine—size, speeds, all that stuff, along with Nick. After that we started interviewing several potential suppliers, and we received machine

quotes, and then we got that package together and submitted it to Tom and to management for approval and uhmm . . . along with a cost analysis, a financial analysis that we got from the business unit. And then we selected a supplier and started working with him. . . .

Project members see it in the same way. For instance, here is the account of a technician on *Delta* on the cooperation between Fred and Nick and the role it played in the successful completion of the project:

> He [Nick] works with purchasing, identifies the right machine manufacturer, schedules all the equipment and the technicians, maintains the machines here [in the laboratory]. . . . Fred defines the technical specifications for the machine . . . like speed, size . . . these things. . . . They work very well together because they understand each other. They speak the same language [technical]. Nick has a good machinist background and they talk. . . . Fred makes all the technical decisions, and Nick finds the right people, the right equipment, and keeps the project on budget. It is a perfect arrangement.

The fact that they "speak the same language" and understand one another and yet have different areas of expertise makes this relationship complementary, as opposed to competitive. They hold one another in high regard for their respective contributions to the project. This, in turn, commands the respect and cooperation of the project members. There is evidence for this in the account provided by a project member on *Gamma* in which he describes the professional "cohabitation" between Nick and Fred and its effect on group morale:

> Nick is a great guy. You can always rely on him. If you ask him something and he doesn't know it . . . he will tell you so, and then he will find the answer for you. But . . . he knows everything about this organization and all the technicians here. Fred is a technical genius. A terrific person! He didn't want to be a manager. He likes mentoring. He left his previous company because they promoted him to management and he wanted to be an engineer and to do design work. . . . Nick makes his life easy, because he [Nick] takes care of the stuff Fred doesn't like doing. . . . They are great to work for.

The critical importance of having both competencies present on a project is further illustrated by the project structure of *Alpha,* a "low success" project that had not been assigned a managerial "star" member. In addition, *Alpha's*

project manager, Fred, by his own account dislikes managerial duties. He did indeed leave the company he worked for prior to joining Global East when he was promoted to a managerial position. To him, that meant less engineering design work, very little mentoring, which he loves, and a lot of paperwork. He accepted the position they offered him at Global East only after he and the laboratory director agreed that even if Fred's official job title might be "project or group manager," he would "be doing the engineering design work" and training young scientists and engineers. As the lab manager concluded when we discussed *Alpha*, "that project needed someone like Nick or Olga to keep the timeline straight."

I bring to a close the discussion on the contribution of technical and organization skills to the success of innovative projects with a quote from my interview with a group manager. Her account is self-explanatory and to the point. In it, she aptly summarizes the points brought so far while offering her explanation of what were the two critical factors that made *Gamma* the smashing success that it was. Let us examine what she has to say:

> The success on this project was based on a balance between the top engineers and high and diverse skills of the technicians. There are two things you worry about on a project: first, to find the right people at the right time with the right skills, and, second, to keep the project within budget. We were very successful on both.

Critical Success Factor 4: Strong and Sustained Support from the Company's Corporate Management

"Strong" corporate support is manifested through the political, financial, and personal commitment to a project's success from a high-ranking company officer(s). What is more, such a commitment, the interview data suggest, must be *sustained* throughout all project-development phases until the project's completion. The advantageous effects of having such support are understood by the organization's members through the themes of securing financial support for an ongoing project, a way to "guarantee" a project's technical success, and a strategy to maintain and enrich the laboratory's excellent technical reputation by continuing to work on highly visible projects. Evidence for the direct connection between strong and sustained corporate support and a project's

outcome is abundant in all the respondents' narratives. The laboratory director, Tom, expresses it unambiguously:

> Success is also political, not just technical. You understand what I am saying. . . .
> On this project, for instance [points to *Zeta*], we started with a very small budget.
> Now, because of the corporate commitment to it, we got a budget ten times higher
> [than three years ago]. . . . Even when you are dealing with technology you cannot
> get it through [the project] unless you take it successfully through the internal
> politics. . . .

Here Tom sees the project's technical success as an outcome resulting from a process that starts with securing the needed political support, and, through it, financial support for the project. He elaborates further: "The higher budget you have, the more things you can try, the more new things you can learn, and the facility gets more visible." Thus, he sees financial and political backing as almost inseparable. The director is so adamant about the critical role of the corporate support that one of the criteria he uses to predict a project's success is "the number of VPs [vice presidents] interested in it." The unfolding logic, as he describes it, is as follows: strong corporate support is expressed in financial support that, when put to use by technically competent people within structural arrangements that are reflective of the nature of the process, is likely to result in technical success. The latter then, in its turn, will strengthen the reputation of the laboratory, and consequently will mean a higher likelihood of future corporate support for new projects. Thus, the lab director sees the contribution of this factor in its ability to trigger a spiral of organizational responses that carry the potential (all other things being equal) to result in a successful outcome. The spiral will reproduce itself on every project so long as the triggering mechanism is present—that is, corporate support for a project or program is secured. The account of one of *Beta*'s managers of what best explains this project's success echoes Tom's sentiment:

> The divisional VPs were ready to back us no matter what because they wanted
> something new, different, something that is a showcase, where they can bring
> customers, [something] that is a high-tech product, a high-tech image and had
> a high-tech manufacturing facility. . . . And for us, [another benefit was] to be at
> the interface and to be constantly in the spotlight. . . . There were a lot of people
> looking at us. This is very good for the facility.

Olga, who characterizes herself as a technical "geek" and enjoys working on stimulating and demanding projects, adds another dimension to the benefit of strong corporate support. She sees in the attention of the higher-ups a promise for more interesting work in the future and draws a benefit from it not just for people like her, but for the organization as well. Here is how she puts these elements together: "This is very good for the [name] facility. . . . This is how we get high profile projects and not just *work.* . . . [her emphasis]." Ted, one of *Zeta*'s project managers, who is customarily laconic, describes the positive effect of Tom's organizational and political skills to the success of his project in the following way: "Tom got the VPs interested in what we do and this way we got a lot of support for developing the program."

The evidence of the instrumentality of this factor is even stronger and more prominent when clear-cut failing projects are discussed. Tom's account of two of the failed projects at the facility, which he brought up in the discussion as examples, describes the process unequivocally.[11] He attributes their outcomes to "political failure" and a change in the company's strategy:

> We had projects that never saw the light. . . . When we stopped working on them, we did not have a solution, so in this regard they were failures . . . for a number of reasons. Time and money was spent, goodwill on the part of management was lost. For instance, several years ago the company's senior management changed. New people were brought in, new criteria for success were established. We had a project that made perfect sense under the previous management strategy, but did not under the new strategy, and that is why it did not come to fruition. . . . The investment decisions of the company have changed. . . . There was no longer a clearly identifiable business unit that was pulling for it, championing it. So, it was simply a matter of budget. There was no justification of putting into it an investment of X amount of dollars when there was no business unit pulling for it.

The director's narrative makes a lot of sense given that it is an industry practice that a division's vice president has at his or her disposal "a pot of gold," also known as a budget, which the VP distributes among the division's respective laboratories and the projects he or she supports for development. A lack of corporate support, as suggested earlier, translates into a lack of funding for the projects an R&D lab would like to undertake. This is the stage of the process at which Tom and Olga's political and organization skills are put into

use. The battle for the attention of the appropriate corporate decision makers is eminently tough, as it is well-known that it could not be taken for granted on the basis of the existing formal reporting relationships. Just because there is a direct line between an R&D director and a company's vice president drawn on the formal organizational chart (in other words, a reporting relationship is prescribed) does not mean that the VP is likely to give each of the directors who report to him or her undivided and equal attention. This is true even more so in the context of a large multinational company with dozens of R&D facilities spread across the globe and more than two thousand researchers and engineers fighting for the attention of a finite number of decision makers—the company's president and vice presidents. To make full use of this formally prescribed relationship, each director must use his or her social and political skills not simply to get, but to *sustain*, his or her superior's attention and interest until the project is successfully completed. In other words, an R&D director needs to use all of the available social capital to secure the top management's uninterrupted support throughout the project-development process.

The researchers, engineers, and technicians too clearly recognize and appreciate the political skills of their managers, Tom and Olga's in particular. Their collective sentiments are expressed in this researcher's statement:

Tom really knows how this company works. He gets their attention [corporate VPs] and gets projects that are going to get us good exposure because they are going to have a good impact on the business. Then he puts together a team of people who can work together, people who have respect for each other and for the work they do.

A researcher's comments on the relationship between the type of projects the R&D lab gets and Tom's political and organizational abilities reveal no discrepancies between the views held by managers and those by the project members:

We work on interesting and challenging projects. The boss [Tom] is very competent technically and politically. He knows his stuff. . . . He has a ton of ideas. . . . He knows what projects will fly and how to make them fly. He knows everybody that is somebody in the company, and these people talk to him. . . .

How critically important it is to have the support of a high-ranking corpo-

rate official for a project or a program's success could be further demonstrated by looking at Global East's decision-making process. Decisions concerning the viability of each project are made on a continuous basis as the work on the projects goes on. There are two decision-making levels that affect the viability of a project—those of the laboratory directors and the divisional vice presidents. Technical decisions and decisions based on the probability of a project's success are made by the laboratory directors; budgetary decisions, however, including the ones about a specific project, are based on the overall budget size for the division and are made at the VPs' level. Without their support, the life of a project could easily be cut short and vice versa. Therefore, a laboratory director must keep the VPs of the relevant divisions interested throughout the duration of a project and even beyond. Obviously, this requires a specific set of skills—social, technical, and political—as there are at least twelve other R&D directors within the company who pursue the same agenda. Here are some examples of how Tom's skills paid off.

Gamma was "blessed by the president of the company and the division's vice president." There was so much faith in the potential business impact of this project that no strict budgetary constraints were imposed on it. The projected budget for *Gamma* was twice as high as the average for the laboratory, while the actual figure went 10 percent above the projections. The division's VP was even listed on the project as a sponsor and involved, to a degree, in the project's decision-making process. The project resulted in a spectacular success. The technological innovation that came as a result of the project introduced a significant change in the way the industry operated, and the company captured about 90 percent of the respective market.

Beta was no different. Two vice presidents were committed to this project's success, and "money was not an object." The result—*Beta*'s outcome exceeded the initial financial expectations fourfold. The projected budget for *Delta,* on the other hand, was low, while the actual cost exceeded it nearly sevenfold. Without the support of the VP of business, *Delta* would have been cancelled and the new business that it created for the company would have not materialized. *Zeta*'s history paints no different picture. The project started with what a team member called an "unusually small budget." Three years later, after the respective division's VPs became unreservedly convinced

of the project's technical and business merits, their commitment was reflected in *Zeta*'s budget, which grew ten times higher than the original. Tom left nothing to chance and co-led this project with Ted, a young engineer. Ted, who by his own account had a real-time "lesson on organizational politics," credits the success of the project to Tom's political skills:

> The success was technical and organizational. Technically, a lot of time you tell somebody a theory, but it sticks better if you can show that it really works. So the technical success was taking the theory and then proving that it really works. Coming up with a process to do that. But we could do it because Tom got the VPs interested in what we do, and this way we got a lot of support for developing the program. . . .

Alpha's "low success" outcome, on the other hand, was largely explained as a consequence of the fact that no high-ranking corporate officer took a direct interest in the project, which translated into a lack of time and financial support needed for additional testing.

Not only do these examples demonstrate the critical role of securing the support of corporate management for a project's success, they also reveal its source—the high reputation of the laboratory's technical and managerial "stars" and their record of success; the stockpile of human and social capital.

THE EFFECT OF THE FOUR CRITICAL SUCCESS FACTORS: SEPARATE OR TOGETHER?

Going back to the literature on innovation, product development, and project management reveals that open communication, strong support from corporate management, and "stars" being assigned to projects have been identified in various, yet separate, streams of prior empirical investigations as critical to technical success, among many others (Brown and Eisenhardt 1995). The findings on their positive effect though, have been inconsistent and, at times, even contradictory. For instance, while Rothwell (1992) and Page (1993) found top-management support to be a prerequisite for success, Rubenstein, Chakrabarti, O'Keefe, Souder, and Young (1976) concluded that excessive support has negative consequences. Yet Kleinschmidt and Cooper (1995) argued that support from top management contributes to success as often and as much as it does to failure on projects. Open and unrestricted communica-

tion, Brown and Eisenhardt (1997) argued, benefits some projects but not others. Others showed that the relationship between the frequency and duration of communication and the performance of new product development teams is curvilinear (Patrashkova and Comb 2004). In the same vein, both scholars and managers have long been aware that having technical and managerial "stars" on projects, although a must, does not guarantee success, as "stars" are also known for having a tendency to be arrogant and "to suffer from the 'prima donna' complex." This could potentially, and often enough does, create a negative work environment as a result of competing and battling egos. The phenomenon is well documented for organizations that depend heavily on creativity. More important, the existing literature on teams, small groups, and project management has demonstrated that none of these four conditions can shape the outcomes of technical projects on its own. Thus, it is the nature of the relationship between these factors, as the analysis of my qualitative data reveals, that adds new knowledge about the dynamics of project management.

A recurring observation that goes through the interviewees' narratives is that the project members do not see these four factors as producing the outcome separately. The narratives of the projects' members and managers tell vivid analytical stories not just about the crucial importance of any one of the four conditions, but also about the interrelated nature of their effect on the outcome. The lab director, for instance, offered elaborate accounts of high and low success on each project. Tom often talks about understanding success from a "system's point of view." By the latter, he does not simply mean the identification and the mechanical implementation of the factors necessary for a project's success. Rather, what he considers to be of greater importance is whether these factors are put together in such a way that they "interact to the benefit of the project." Perhaps as a reflection of his high position in the formal structure he clearly assigns prominence to the political aspect in this endeavor, expressed in securing corporate support for each project:

> Your effectiveness is judged based on how much they [your customers] ask for your product and the services you provide. . . . Your success also depends upon the perception of your customer [the business unit] about the people on the project, how skilled they are. For example, on these projects, if the business unit did not

perceive this person [points to a project member from the lab] to be a valuable person, our contribution goes down, it is not as high as it could be. . . .

Thus, Tom argues that technological success is a result of several factors acting together. He draws a direct link between the reputation the laboratory has across the company with regard to the quality of its technical expertise and the likelihood of securing strong corporate support for the lab's programs. His narratives on each of the six projects incorporate a multitude of factors and circumstances that interactively shape the outcomes. The organization and project members, too, tend to see these four elements producing the outcomes of the projects in interaction; each one as a prerequisite for the others.

I triangulated the findings derived from the inductive analysis of the qualitative interviews on both the four critical success factors and their interactive effect with the results from the application of Ragin's methodology, Qualitative Comparative Analysis (QCA) (2000, 1994, 1987).[12] Ragin's method allows us to test complex models while addressing a major issue in social science research—namely, it is hardly ever the case that one or several factors independently provide a satisfactory answer to the question of interest. Moreover, the wide tapestry of social conditions does not preclude the possibility of different combinations of causal conditions leading to the same outcome and, thus, being equally valid explanations. This comparative approach is "especially well-suited for addressing questions about outcomes resulting from multiple and conjectural causes—where different conditions combine in different and sometimes contradictory ways to produce the same or similar outcomes" (Ragin 1987: x). The method is applicable to examining patterns of similarities and differences within a moderate number of cases and is appropriate for small-scale studies.

The final result from running the QCA is expressed in the following equation:

$$S = f \cdot C \cdot T \cdot O.^{13}$$

It should read as follows: a project that is characterized by a low degree of formal communication (f), *and* a high degree of corporate support (C), *and* on which there is a person(s) who occupies a highly central position in the technical-advice network (T), *and* on which there is a person(s) who occupies

a highly central position in the organizational-advice network (O), "more often than not" will result in a "high success" outcome (S). The results from the QCA show that the *necessary* and *sufficient* condition for the projects' "high success" outcome is the combination of these four factors acting in conjunction, as opposed to the same four factors acting independently and still being able to produce the same result.

Hence, the sheer presence of each condition will not be sufficient in and of itself to shape the success of R&D projects. Furthermore, in the respondents' narratives, they seem to be entwined to a degree that makes it difficult to separate their effects through the social structures. For instance, as it was argued by virtually all lab members, sustained corporate support functions through the formally arranged decision-making channels. It does appear though, that the relation prescribed through the formal reporting structure is activated and utilized to a large degree by the work of the soft, informal structure through high status and reputation. The open, free, and uninhibited communication patterns do not take place by serendipity; they are sustained with the blessing of the formal structure and the cultural norms at the laboratory that reinforce the structural prescriptions. These communication patterns do not work by circumventing the prescribed flow of knowledge and information; they *are* the formally prescribed pattern. Similarly, the laboratory management seems to utilize the people who are at the center of the knowledge and information hubs by embedding them into the formal structure of the projects, as opposed to letting them compensate for any inefficiencies of the formal structure on their own and outside of the web of formal entanglements. Hence, pointing out where the effect of the formal structure stops and where the influence of the informal begins is not a trivial undertaking, as the two structures are interwoven.

The question then is, *How* does this take place? In the next chapter, I look at the formal and social network structures at the laboratory and on the six projects, and uncover a unique project design mechanism through which the four factors and the two structures are placed into an interactive relationship to the great benefit of creating an environment conducive to perpetual learning and sharing of knowledge.

BEYOND THE FORMAL AND INFORMAL DIVIDE

> "What we need are theories that focus on how the formal and informal structures are interrelated and influence each other. It is important that we start viewing structure not from a design perspective (in which structure is primarily seen to constrain action) but from an action perspective (in which structure is viewed both as the basis for action and as the trace of action)."
>
> *Nohria and Gulati 1994: 529*

THE FORMAL STRUCTURE AT GLOBAL EAST

The Global East R&D laboratory has a rather simple formal structure. It is led by a director, Tom, and consists of three groups each headed by a group manager—Fred, Olga, and Nick. The people who work full time in the facility are listed under the group managers, starting with the senior researchers and senior engineers, followed by junior researchers and engineers, after whom are listed the trainees and the technicians. Besides the director and the three group managers, though, there are no strictly demarcated hierarchical lines. In general, all laboratory members are directly involved in projects, regardless of whether they hold a managerial position or not. Individuals are assigned to projects based on their expertise and as needs arise. Experts from other business divisions are often brought in as project members, and outside consultants are hired if and when needed. It is a common practice that one works on several projects simultaneously, carried out by any of the three groups. The atmosphere within the laboratory is quite informal, and people refer to one another by their first names as opposed to using their titles. Figure 3.1 depicts visually the laboratory's formal structure. Those organizational members who played a significant role on one or more of the six projects under study, and to whom I refer later in the book, are marked by their pseudonyms; the rest are identified by a letter N and a corresponding number.

THE INFORMAL STRUCTURE AT GLOBAL EAST

As is often the case, after an R&D manager explains to an outsider the reporting relationships and the roles and responsibilities allocated to individuals, as depicted by the organization's formal chart, the explanation is followed by the remark, "but this is not how this organization *really* works. . . ." To find out how it *really* works I examined the social network structure in the laboratory—

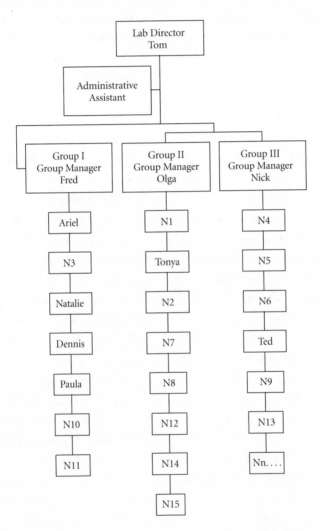

Figure 3.1   Formal organization chart: Global East's R&D laboratory

those stable patterns of informal social relations that the members enter into in the daily process of conducting technical research and development.

The R&D laboratory, as I mentioned in the previous chapter, was established about ten years prior to this research investigation. Throughout this period of time, turnover has been very low; about 80 percent of the people who joined the organization back then remained working there at the time of the study. In addition, lab members have overlapping project assignments, and thus, virtually every scientist, engineer, and technician worked, at some point, on a project with every other one. Naturally, this arrangement provides ample opportunities for people to get to know each other well, both as professionals and as personalities, and to interact with one another in myriad ways, thereby developing various social relationships.

That social networks are not only multifaceted but also operate on different levels depending on the type of relations that individuals maintain has been previously established (Burt 1983; Tichy, Tushman, and Fombrun 1979). In this study, to understand how the laboratory actually works, I was concerned not only with the effect of different types of social networks on R&D projects but also with examining the content of the information and knowledge exchange that is specific to technical environments. To capture these multiple social ties, I asked each laboratory member questions about the type and quality of their interactions with co-workers in the R&D organization.[1]

The vital importance of gathering and transmitting technical information to innovative systems has been well researched and documented by Allen (1977, 1984), Ebadi and Utterback (1984), and Tushman (1977, 1978). Furthermore, the research on boundary spanning has demonstrated that the informal channels are critical to the effectiveness of R&D organizations in disseminating the technical knowledge across "different communication areas"; for instance, intralaboratory, intraorganizational, and extraorganizational (Ancona and Caldwell 1992; Tushman and Scanlan 1981). In my preliminary interviews, however, the researchers, engineers, and technicians indicated that it was not just information that they were seeking in their daily work; they were very much in need of advice. Furthermore, advice on technical matters to them was different from advice on organizational matters (such as scheduling, budgeting, assignments, and coordination). As I reported in the previous

chapter, they also saw some lab members as being technical "stars" and others as managerial "stars." Hence, I constructed two separate work-related advice networks—*technical* and *organizational*—to reflect the different types of advice that people sought while working on R&D projects (Rizova 2002, 2006a). To measure the technical-advice network, I asked the lab members to whom they would go if they "need someone with technical competence and skills." Conversely, to measure the organizational-advice network, the question asked who they would contact if they "need someone with organizational competence and skills." Furthermore, I asked the lab members to elaborate on *why* they choose to seek advice from a particular person.

In addition, two other social networks were also examined—*instrumental* and *expressive*. The instrumental ties focus on the relations between individuals that concern work-related content, such as the transfer of physical, informational, or financial resources (Ibarra 1993; Tichy, Tushman, and Fombrun 1979). Hence, these ties present a broader view of how work-related content is more generally exchanged, in that they differ from a technical-advice network that maps the relationships that people tap to solve specific problems. To capture the instrumental ties I asked my respondents to identify the co-workers with whom they "primarily talked about problems that arose in the course of a project." The expressive ties, on the other hand, concern friendship and social support. Such relations are not prescribed by the organization's formal structure, or rules and procedures, as they do not involve work-related attitudes and behaviors (Ibarra 1993; Gibbons 2004; Mehra, Kilduff, and Brass 2001). To tap those ties, I asked the lab members to name the people from the R&D unit with whom they socialize outside work, such as going out for a drink or spending time on weekends, and felt comfortable discussing what is going on in the organization in general. I also asked respondents to elaborate and provide specific examples in their answers to each of the four social network questions. As a result, I examined the following four social networks: instrumental, expressive, technical-advice, and organizational-advice.

To identify the "hubs" of social relations in each social network I computed in-degree centrality for the forty-two people who were employed full time in the R&D laboratory.[2] In-degree is a simple, yet powerful, measure that focuses on the relative position that individuals occupy within a social network; it

is a well-established measure of communication activity (Burt 1980; Marsden 1990) and social status (Wasserman and Faust 1994). Hence, if a network member were to occupy a highly central position in a social network, that would indicate that he or she is "at the center of a number of connections, a point with a great many direct contacts with other points" (Scott 2005: 83). The top part of Table 3.1 provides descriptive statistics for these measures, while the lower part reports the descriptive statistics for the in-degree centralities of the seven laboratory members who have the highest centrality scores in the R&D organization's four social networks. I refer to these individuals by the following pseudonyms: *Fred, Natalie, Nick, Olga, Ted, Tom,* and *Tonya.* Their mean centrality is above the mean centrality for the laboratory, which indicates the highly central position that these seven individuals have in the lab's social network.[3] These lab members have centrality scores that are significantly higher than their co-workers' in all four social networks. This is evident from the aggregate measure of the position of the central individuals across

Table 3.1

Network centrality: Central individuals and other R&D laboratory members

	Descriptive Statistics	Instrumental Network	Expressive Network	Technical-Advice Network	Organizational-Advice Network
Whole Network (WN)	Mean	1.45	0.95	1.55	0.88
	SD	3.04	2.17	2.62	2.38
	Min	0	0	0	0
($N = 42$)	Max	17	10	11	11
Central Individuals (CI)	Mean	6.29	4.29	5.86	4.57
	SD	5.06	3.95	3.93	4.35
	Min	2	0	1	0
($N = 7$)	Max	17	10	11	11
Z score (mean WN– mean CI)	--	1.59	1.53	1.65	1.55
t value, one-tailed test (CI vs. others)	--	6.56***	6 .23***	6.09***	7.08***

NOTE: *** P < 0.001

Table 3.2

Network positions of the individuals central in the laboratory's four social networks

	Tom	Nick	Fred	Tonya	Olga	Natalie	Ted
Instrumental Network Centrality	Medium (5)	Low-medium (4)	High (17)	High (8)	Low-medium (4)	Low-medium (4)	Low (2)
Expressive Network Centrality	Low (2)	High (10)	High (8)	Low (2)	High (7)	Low (1)	Low (0)
Technical-Advice Network Centrality	High (11)	Low (1)	High (11)	Low (2)	Medium (5)	Medium (6)	Medium (5)
Organizational-Advice Network Centrality	High (11)	High (11)	Low (3)	Low (1)	Medium (5)	Low (0)	Low (2)
Total Centrality	*High (29)*	High (26)	High (39)	Low (13)	Medium (21)	Low (11)	Low (9)

the networks (Ibarra 1993), which is presented at the bottom of Table 3.2. The table describes the central positions in the four networks, which are classified as high, medium (that is, about the mean for the group of seven central individuals), or low (that is, below the group's mean).

On the basis of the four network questions, I compiled the instrumental, expressive, and technical- and organizational-advice networks. These networks can be visualized as directed graphs, as presented in Figures 3.2–3.5.[4] The forty-two individuals who worked in the R&D laboratory are presented in the following groupings on each of the four sociograms. The seven lab members with the highest degree of centrality in each network are in the center and are identified by their pseudonyms. Consistent with the lab's practices, most of them worked on multiple projects. For instance, Fred worked on *Alpha, Gamma, Delta,* and *Epsilon.* Tonya was assigned to *Beta* and *Epsilon,* Nick to *Gamma* and *Delta,* and Olga worked on *Beta* and *Zeta.* Tom, who is the laboratory director, worked on *Zeta.* The other two central lab members, Natalie and Ted, worked only on one project each, *Epsilon* and *Zeta,* respectively.

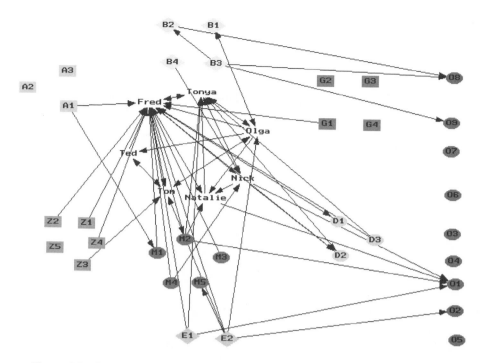

Figure 3.2 Instrumental network

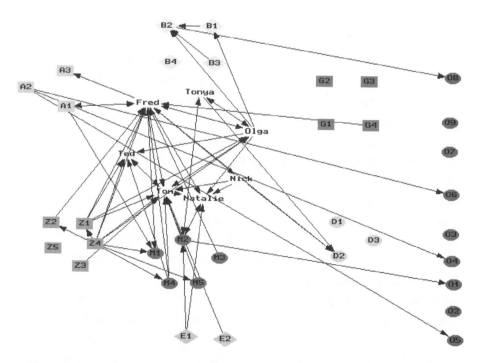

Figure 3.3 Technical-advice network

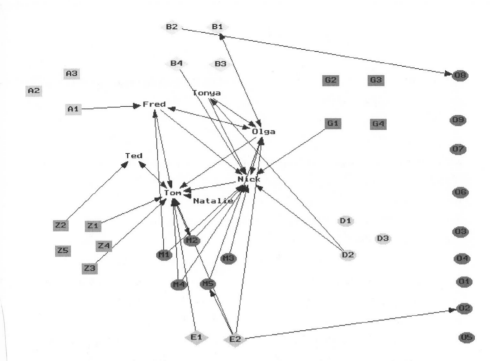

Figure 3.4   Organizational-advice network

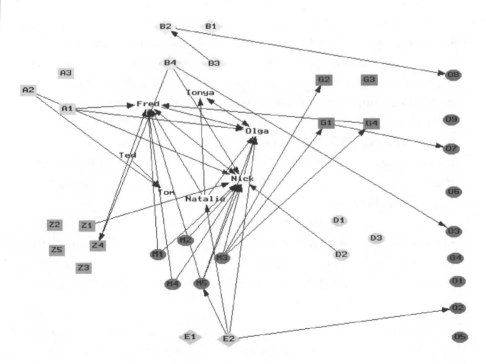

Figure 3.5   Expressive network

The group of social network actors, comprising five people who were assigned to multiple projects at the time of the study, are labeled as M1 through M5. Those researchers who worked on just one of the six projects under investigation are labeled with a letter for their respective project (*Alpha, Beta, Gamma, Delta, Epsilon,* and *Zeta*) and an identifying number—for instance, B3 for a member on project *Beta*, or G1 for *Gamma*. Finally, the last group of nine people, represented on the right-hand side of the sociograms and marked as O1 through O9, are the R&D lab's members who were not assigned to any of the projects I studied.

Further examination of the data in Table 3.2 and the sociograms points to several interesting observations. Although the seven most central members at the R&D laboratory are easily identified, their centralities vary in the different networks. For instance, some have centrality of 0 in individual networks. This is the case with Natalie in the organizational advice network and Ted in the expressive network. Tom, the director, occupies highly central positions in both the technical- and organizational-advice networks, but not in the expressive one. Interestingly enough, the lab members with the highest centralities in this social network are the three group managers Nick, Fred, and Olga. Tonya is central in the instrumental but not in any of the other three social networks. A rather intriguing observation concerns Nick and Fred's centrality scores in the two advice networks—they are almost perfectly inverted. Nick is highly central in the organizational-advice network but not in the technical-advice network, and Fred's standing is just the opposite. With both being not just managers on many projects but group managers as well, I had expected that they would be approached for advice on both technical and organizational matters (Brass 1984; Krackhardt 1990). These are also the two networks the respondents offered the most elaborate responses to.

Going back to the qualitative data, and examining the accounts of the lab members, sheds some light on this situation. When asked to elaborate on their answers as to why they named Nick as the person they go to for advice on organizational matters, the view of the majority of the scientists is captured in the account of their colleague who says,

> As I said earlier, he knows how to get things done around here. That's one thing. But he is easy to talk to too. Going to him with a problem or a question does not

make you worry that this will blow up in your face. He is just a very decent guy; very trustworthy.

In his account, the respondent draws a link between Nick's human capital and social capital; he sees both as necessary conditions. A technician, too, explains his choice by pointing to the same connection. He states, "You can talk to him about anything. He will keep it to himself and will not use it against you sometime down the road. . . . No backstabbing . . ." His sentiment reappears later when the same interviewee explains why he likes socializing with Nick outside of work:

> Most people think techies are these boring nerds, but you need to be around these guys for a day to know. Nick is just hilarious . . . and very easy-going. He does not have the "I got you" mentality. Whatever is said to him, stays with him no matter the circumstances. . . . There is a lot of trust.

Hence, being a competent professional is not enough of a reason for the people at the laboratory to turn to a person and ask for his or her advice on "the latest mishap." He or she must also be someone who can be trusted, someone who is not seen as malicious and will not use knowledge about or from a colleague that is gained in one of those less-guarded moments to his or her own advantage at a later point.

Interestingly enough, the very same reasons, a combination of both human and social capital, were offered uniformly by all eleven people who cited Fred as their best source of advice on technical matters. Here is how a young engineer explains his preference:

> Well, it isn't just that he loves what he does and has a lot of experience. It gives him pleasure to solve a problem; he loves it and he does not make you feel stupid for asking. The guy is a genius with a ton of experience. . . . you can learn a lot from him. I am sure you have heard that a lot. He never makes you feel incompetent though. . . . He wants us to learn. I was a trainee here first, and he was my mentor. Other people with his kind of experience and track record will easily intimidate you, but he makes you comfortable. He is happy when you learn, and no question is small enough to ask him. And he'll tell you that he learns from you too.

The shared sentiment that just having the technical expertise does not make one a good person to seek advice from is even more poignantly articulated by

a technician who goes through the trouble of naming a senior lab member to whom he would not go for advice and the reasons as to why not:

> You see, there are people with technical experience and there are those that have the people skills too. Fred has both and this is why I go to him for advice. He is a problem solver . . . cares about finding a solution . . . does not make you feel bad for asking for help . . . A problem solved makes his day. See, I will not go to him [points to Ariel on the organizational chart]. He is a senior engineer too, and he knows what he does. He has a lot of experience and is very good at what he does. But . . . sort of treats you like a schoolboy . . . the way he asks you what you have tried so far and so on, tells you that he doesn't think much of you. He just has to show you that he is better than you. . . . Anyway, that's why I don't go him. I'd rather ask Fred and Olga. I will go to Tom too; he is the boss and he is very busy but he wouldn't mind. . . . He will take the time to talk to you because he likes to know what issues we face.

Being socially apt and a knowledgeable colleague are the two themes that have been highlighted repeatedly by virtually every single person that I interviewed on the matter of seeking advice. These same themes came through when the organization and project members discussed the work atmosphere in the facility and the ways in which it contributes to the success on projects. They are echoed again in the account offered by an engineer about Fred's centrality in the technical-advice network.

> If there are any technical issues or doubts, I go see Fred. . . . If he can't help, he knows who to send you to and who else to ask outside of the lab. . . . He is just this really, really smart guy when it comes to this, and he is really someone who you can trust. So if you really have a problem or just need to double-check your ideas, everybody says—go and see Fred and he would straighten you up. Even if I just need reassurance . . . , I go to him. . . . It is my feeling anyway. He has been doing the same thing for many years now and he knows a lot about it. . . . Very intelligent, very friendly, very willing to work with the newer, younger guys and to show them how to do the job correctly.

Here, the engineer too emphasized that both specialized knowledge and trust are the necessary preconditions for seeking work-related advice from a colleague. Trust that one can ask any question; trust that this will not be construed as "incompetence"; trust that one's vulnerability will not be used

against him or her when the time will come for a performance review; trust that this is to the benefit of all and to solving a problem.

What is even more interesting is that trust multiplies; by being encouraged to seek each other's help, the lab members become aware of their work and expertise, and new advice "stars" emerge. For instance, this is how a senior engineer explains why one of the people he seeks technical advice from is Natalie.

> She [Natalie] was a trainee first for a year and then she got hired. She kept an open mind and was very eager to learn. She wouldn't shy away from anything, very thoughtful and hands-on person. So, I got to know her because she was curious and was asking questions. . . . This is how I came to know how much insight and knowledge she's got. She established herself very quickly as a person with solid technical background and expertise. . . . She is great to run things by, to point your thoughts in a direction, to . . . give you a new insight because she is not afraid to be wrong. A very pleasant colleague too.

Furthermore, seniority in the formal structure or tenure with the organization do not seem to bear an association with one's centrality in the laboratory's work-related advice networks. Of the seven most central members in the laboratory's social networks, only Tom and Nick have worked at the facility since its establishment. Fred has been with the organization for eight years, Ted and Natalie for about five, and Olga for three and half years. Tonya joined the lab just about two years prior to the study. She was hired because of her very specialized technical skills in two distinct areas. Prior to co-managing *Beta,* she had worked on only one other project that was closely related to one of her areas of expertise. Her performance on it gained her the reputation of a "problem solver," and she came out central in the lab's instrumental network. As to their positions in the formal structure, Tom and Olga are the two out of the seven central individuals who began their careers at the lab occupying the top formal positions they currently hold—a director and a group manager, respectively. The other five members were either promoted after working in the laboratory for several years, Fred and Nick for example, or continue to hold the formal positions they were hired at.

These observations raise interesting questions about the ways in which the R&D laboratory at Global East draws on the knowledge and expertise that resides in its most prominent members and puts into use their social capital.

PROJECT DESIGN FOR R&D SUCCESS

To explore further the formal-informal structural interaction, I looked at the project-design practices as well as at the assignment of roles and responsibilities at the R&D lab. To this end, I mapped the findings from both the qualitative and the social network analyses onto the formal structures of each of the six projects. First, I checked whether each of the four conditions found to be indispensable to the success on R&D projects was present on the six projects that I studied. This step included checking whether individuals who occupy positions of centrality in either or both of the laboratory's work-related advice networks, in other words, *technical* and *organizational* "stars," had been assigned to the projects under investigation. The findings from this step are reported in Table 3.3. As they show, on the five "high success" projects all four critical success conditions were met, while the one "low success" project, *Alpha,* lacked two of them—corporate support and not having been assigned an individual with managerial "star" qualities.

Next, I matched the individuals who held positions of centrality in the technical- and organizational-advice networks to the positions that they held in the formal structure of each project as well as to the roles and responsibilities they were assigned on these projects. In their narratives, as I discussed in Chapter 2,

Table 3.3

Presence or absence of the four critical success factors, by project

Project	Low Degree of Formal Reporting Communication on the Project	Individual(s) Central in the Lab's Technical-Advice Network Is on Project	Individual(s) Central in the Lab's Organizational-Advice Network Is on Project	High Degree of Corporate Support for Project	Project's Outcome
Alpha	Y	Y	NO	NO	Low success
Beta	Y	Y	Y	Y	High success
Gamma	Y	Y	Y	Y	High success
Delta	Y	Y	Y	Y	High success
Epsilon	Y	Y	Y	Y	High success
Zeta	Y	Y	Y	Y	High success

the respondents did not directly link the instrumental and expressive relations to their understanding of how and why projects succeed. This finding was also triangulated with the results from the Qualitative Comparative Analysis (QCA) (Ragin 2000). As I could not map the results from these two social networks on the findings from the qualitative inductive data analysis, I did not include them in the further analysis. Last, I went back to the interview transcripts to look for overlaps or discrepancies between the roles and responsibilities that these project members were formally *assigned* on the projects they participated in and the ones they *actually* performed. In other words, I wanted to see to what degree there was a discrepancy or correspondence between the *prescribed* and the *actual* decision-making authority.

This analytical procedure allowed me to capture the scope as well as the degree of the contribution of these individuals in shaping the outcomes of their respective projects. It also illuminated the project-design processes at the R&D laboratory. The results from this matching process are reported in Table 3.4. In it, for each project, I list the individuals who occupy positions of centrality in both the projects' formal and social network structures. I also report their centrality scores in the two advice networks—*technical* and *organizational*. As the six projects differ in size—for instance *Alpha* had five project members, whereas *Epsilon* had ten and *Zeta* twelve—I computed normalized in-degree centrality scores to make the comparison between projects meaningful (Borgatti, Everett, and Freeman 2002).

The data from the table, in conjunction with an examination of the qualitative interviews, reveal that on the "highly successful" projects those who occupy positions of high centrality in the lab's technical-advice network also occupy positions of high centrality in both their projects' technical-advice network and formal structure—they are the projects' managers. The data from the interview transcripts confirm that their areas of responsibility cover technical matters. By the same token, the individuals who are central in the laboratory's organizational-advice network occupy positions of high centrality in both their project's organizational-advice network and formal structure. These individuals have been assigned to the projects either as co-managers or as members, and their areas of responsibility cover managerial logistics, such as budget and time management, as well as the coordination of tasks and

Table 3.4

Central individuals in the formal and the advice network structures, by project

	Formal Structure		Normalized In-degree Centrality in the Advice Social Networks		
Project	Project Manager(s)	Centrality in the Technical-Advice Network (occupants and scores)	Mean and Standard Deviation Scores	Centrality in the Organizational-Advice Network (occupants and scores)	Mean and Standard Deviation Scores
Alpha	Fred	Fred (50.0) Ariel (50.0)	Mean = 30.0 SD = 18.7	None	NA
Beta	Tonya and Olga	Olga (50.0)	Mean = 19.0 SD = 16.4	Olga (33.3)	Mean = 7.1 SD = 12.1
Gamma	Fred and Nick	Fred (57.1)	Mean = 8.9 SD = 18.8	Nick (57.1)	Mean = 4.1 SD = 18.8
Delta	Fred and Nick	Fred (33.3) Dennis (50.0)	Mean = 14.2 SD= 18.7	Nick (66.6)	Mean = 11.9 SD = 23.0
Epsilon	Natalie	Natalie (71.4)	Mean = 19.6 SD = 22.5	Nick (42.8) Olga (42.8)	Mean = 14.2 SD = 17.4
Zeta	Ted and Tom	Ted (45.4) Tom (72.7)	Mean = 15.9 SD = 21.3	Tom (54.5)	Mean = 7.5 SD = 15.2

operations. This, however, is not the case on *Alpha,* the one "less successful" project, on which a *technically* central member, Fred, is given all the managerial responsibilities, while no *organizationally* central individual has even been assigned to the project.

A rather interesting observation of the organization of the work processes in the R&D laboratory is that some projects have been assigned two managers. The narratives suggest that the number of managers on each project might be a function of the project's technical and managerial complexity, and the availability of lab members who possess the necessary human and social capital. This could also vary by project stage. As expected, and as shown in the previous chapter, managing a technologically innovative project requires both technical and organizational competencies and skills. Consequently, if within the R&D laboratory there is a person whose level of technical skills, expertise, and experience match the project's technical nature and goal, that person is

assigned to the project as a technical manager regardless of his or her position in the lab's formal structure. If the same person is recognized as a "managerial star" as well, or the project's technical specifications and scope only require "routine management," the overall responsibility for the project in question will rest with one manager. This was the case on *Beta*. During this project's first stage Olga was the manager. During its second stage, once the technical specifications were defined and more or less agreed upon, Tonya was given the responsibility to manage the technical side of the project, while Olga reduced her responsibilities to keeping the work on schedule and taking care of the financial side of it.

On the other hand, on projects characterized by a high degree of technical complexity and ambiguity, if an organizational member's human and social capital do not capture the necessary technical and managerial competencies, then two project manager positions are created. The areas of responsibility for each of the two co-managers overlap with their areas of strength. *Gamma* and *Delta*, for instance, are two cases in point. Fred, who is a "technical star" both within the organization and the projects' technical-advice network, was the technical manager; he was responsible for all the decisions concerning the technical approach, the types of testing, the technical parameters, and so forth. Nick, who is one of the three "managerial stars" in the laboratory and also in the projects' organizational-advice network, was assigned to both projects as a co-manager; his areas of responsibility covered budget and time management, personnel assignments, machine and testing scheduling, and dealing with suppliers. Thus, Fred and Nick's human and social capital complemented one another, as each of them lacks the skills the other possesses while together they satisfy the project management needs. On not so highly technical complex projects for which the technical specification and procedures are less ambiguous, there is one project manager. *Epsilon*, for instance, was described along these lines, and Natalie, who is only central in the technical-advice network, was the manager. Nick and Olga, however, were members on this project helping out with scheduling, coordinating, and budget issues.

Thus, on the five "highly successful" projects, positions of centrality in the two work-related advice networks were converted into positions of centrality in the projects' formal structure. In other words, the analysis suggests that in

Global East's R&D laboratory, roles and responsibilities on projects do not mirror the formal hierarchy in the lab. Rather, they seem to be assigned more in line with a member's centrality in the technical- and organizational-advice networks and one's human and social capital, irrespective of one's position in the laboratory's formal structure. A look at the formal organizational chart, as shown in Figure 3.1, demonstrates this point. For instance, Natalie, a junior scientist, is listed as number three in her Group I after two senior researchers—Ariel and N3—yet she was the sole project manager on *Epsilon*. Tonya, *Beta*'s co-manager, occupies an organizational box number two in her Group II under N1, who is a senior researcher. *Beta*'s other manager was Olga, who is also the group's manager. Ted, an engineer from Group III, is found even further down in the formal hierarchy. He is number four in his group—the last engineer before the trainees and the technicians are listed. Yet he was a co-manager of *Zeta* along with the laboratory director. Fred and Nick, who are managers of Groups I and III, respectively, were assigned to co-manage projects *Gamma* and *Delta*.

In the narrative that follows, the process and the rationale that the director applies to the project-design system at the laboratory is unveiled. In it, Tom ties in the themes of motivation, creativity, and organizational and individual learning to the issues of organizational size, sustainability, and survival.

Innovation is not necessarily linked to the size of the company; it is linked to a particular type of preference; it is a preference for a type of behavior. Promotion opportunities are not too many here because we are a small facility. You see, our structure is virtually nonexistent. People need to stay motivated and to be willing to share their knowledge and information. We have to look for other ways to recognize good work other than a job title. We are a new facility, and we need to think about our future in the long run, how to keep people here. These are all highly motivated and skilled people; if they don't feel that they are getting the recognition and the exposure here, they will leave. . . . Knowledge and experience are very difficult to replace. Experience matters a lot in this business. . . .

His discussion of which roles and responsibilities are attached to the structural position of a project manager and who the people at the facility are who can be its legitimate occupants is rather poignant. In the account that fol-

lows, the director is very explicit about the open-ended way in which the title "project manager" is constructed in the laboratory and the effects of the manner in which it is defined.

> The challenge of the manager is to create each project in its initiation phase with some specificity attached to it. I don't necessarily mean by "a manager" a class of people or a job. As a management component of any project, the manager's role is to facilitate the expansion of the project objective depending on the project strategy. . . . When you look at the whole problem, identify opportunities and facilitate creativity and allow for a certain degree of uncertainty. Uncertainty doesn't have to be counterproductive to profitability. You are hedging your bet in terms of uncertainty by having the input of a larger group of people, which a larger company facilitates by its nature and a smaller company does not, because if the boss of a small company does not have the input he has to make a decision. . . . Whoever does it more we call them innovators. Those innovators don't have to be only engineers or managers; they could be scientists, technicians, trainees. . . .

Here, Tom also offers an insight into one of the most vital issues for any manager of technology—how to reduce uncertainty. The director's solution is to seek for knowledge, experience, and skills across different functions and levels and to capture and expand those through the assignments of varying roles and positions. In his words, "in the context of global economy," the goal is to "turn each individual employee into a 'problem solver.'"

The director adds another factor that complicates the life of a manager of technological innovation—the fact that a healthy project portfolio of a company includes routine technical projects together with those that are radically innovative. This presents a manager with March's classical dilemma about how to balance the "exploration of new technical possibilities" with "the exploitation of old technological certainties" (1991: 71). Here is Tom's account:

> And you will also find that the impact of a program is larger provided that there is a greater degree of innovation. And the impact of the program becomes less if the project objectives are so precisely specified and the certainty of the outcome is determined. That does not promote creativity. But every company needs portfolio projects, which balances a blend of the two. But the challenge of every manager in an innovation group is to facilitate each project to kind of be moved into this direction of more certainty but also creativity. . . . [5]

Such an understanding of the innovation process and its management fits in well with the findings from the qualitative interviews. The fluidity of assigning positions and roles on projects, independent from those fixed by the formal organizational structure, is quite apparent from the statement that follows. The director ends the discussion about the challenges that a manager faces in organizing for innovation by tying two final elements into his account—how to motivate people to operate within such a fluid structure and how to sustain their motivation while creating a perpetual learning organization.

> And then you have to make sure that they take ownership of the project; that is, if the project succeeds, each one of them succeeds. And there is this other thing too; it is not just that you reward people monetarily; that is important, of course, and it is expected and yes, you get the pay raise. But you need to create a situation when they get to be respected for what they do and how they do it, for their skills, for their knowledge, for the novelty in their thinking, and how they put this knowledge to use. . . . People like that are more willing to give more of themselves if they know that they can be respected as knowledgeable. . . . It is easy in this environment to spread the news because everyone works on several projects at a time, we all know each other really well. So, if you come up with something new and we try it and it works and does well on the market, everyone here knows about it, I know about it, my boss (the vice president) knows about it. . . . It's like that. What I am saying is that you don't have to write a memo; people talk about it. So, it becomes important for everyone to want to be successful.

Once again, the functioning of the entire system is linked to the elements of human and social capital. Fred describes the structure on the R&D projects in a similar fashion:

> You know, with some of these titles . . . project managers, project leaders, it is very . . . how to say it . . . I don't want to say it's a loose organization, but our responsibilities aren't really fixed that way. . . . What is unique about this facility is the structure of the organization and the way core competencies and responsibilities are combined. . . .

Thus, through the implementation of this unique project design, one that weaves the human and social capital, which originates from the work-related advice networks, into formal structural positions and roles, the R&D management at Global East has found a way to avoid the Peter Principle (Peter and

Hull 1969). In other words, the embedding of the human and social capital into the projects' formal structures prevents members from being forced into an organizational box that prescribes roles, responsibilities, and behaviors that are inconsistent with their aptitudes and abilities. It achieves structural stability and predictability by creating defined, and yet fluid, networks for aligning experts with organizational challenges and objectives, while creating a learning environment that is self-reproducing. Once a person is recognized as competent and trustworthy, and his or her advice is sought, and when a project comes along that requires this person's skills, that person is likely to be appointed as the project's manager. This not only secures the access to this person's social capital through the formal position in the project hierarchy (Lin 2001) but furthers his or her status, not on this project alone, but in the organization and, by extension, within the company. Even when that person no longer is a project manager, his or her centrality in the advice networks is likely to be solidified, and the organization will continue to draw on the person's social capital because that formal position has given him or her high visibility (Brass 1984; Krackhard 1990).

At the same time, in order for this person to maintain high status in the work-specific advice networks, he or she has to continually learn and to compete, so to speak, for knowledge recognition with the "stars," such as Fred, Nick, Olga, and Tom. On the other hand, converting the informal centrality into formal status acts as a motivational mechanism for the other organizational members to strive to achieve the same level of recognition. The structuring of a project, as prior research has demonstrated, influences the pattern of social networks. I would argue that the assignment practices at the lab contribute to the expansion and multiplication of the informal relations (Saxenian 1988, 1996). In this way, Global East has found an answer to one of the most pressing issues in knowledge-intensive settings—how to manage the informal relations to the benefit of both the individual and the organization while reducing uncertainty. By converting the *actual* relations into *prescribed* ones, the two advice networks that are critical to the work in R&D environments are being reinforced and not thwarted by the formal structure. Thus, it becomes a self-reproducing and re-enforcing cycle of capturing, sharing, and using knowledge, as depicted in Figure 3.6.

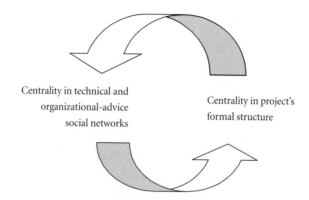

Centrality in technical and
organizational-advice
social networks

Centrality in project's
formal structure

Figure 3.6   A cycle of knowledge transfer

THE EFFECTS OF INTERWEAVING THE FORMAL AND INFORMAL STRUCTURES

The management practices at Global East suggest that success on technological projects is fostered when the formal and informal structures are entwined through the blending of positions and roles on each dimension. This is accomplished not simply by having centrality in the work-related social advice networks recognized, but by converting it into centrality in the projects' formal structures. In doing so, the human capital (the technical and organizational competencies, knowledge, and skills) and the social capital (the relations among members and the trust and reputation they create) (Becker 1964; Lin 2001) are interwoven into the projects' formal structures and performance is enhanced. That "performance is better when communication structure matches the information-processing requirements of a task" (Brass, Galaskiewicz, Greve, and Tsai 2004: 799) has been already demonstrated by Brass as well as by Roberts and O'Reilly (1979). The project-design and role-assignment practices at Global East attend to this matching need.

Thus, if Fred, who is central in the *technical* advice network, is assigned to a project that requires his particular area of technical expertise but not as a project manager, these findings would suggest that his skills would be underutilized, as he would not be formally assigned the responsibility for the technical decisions on that project. Support for this conclusion is offered by Ibarra's

1993 study of the relative impact of individual attributes, formal position, and social network centrality on the exercise of individual power. She found that personal sources of power are of great importance in determining the role one plays in technical innovation (not to be confused with technological innovation).[6] As to centrality in social networks, her results show that "centrality was the most significant predictor of administrative innovation roles," whereas involvement in technical innovation "appeared to be equally responsive to formally and informally derived sources of power" (Ibarra 1993: 492).

On the successful projects in Global East's R&D lab, roles and responsibilities match one's centrality in the technical- and organizational-advice networks. In this manner, the human and the social capital, and by extension the financial capital, are not only behaviorally but also structurally embedded. As a result, the resources of the organization are easily accessible and readily mobilized as there is no separation between the assets created and leveraged through the *prescribed* relationships and those generated by the *actual* relationships. Such a mode of organization allows the managers at Global East to capitalize on the benefits that the social networks, together with the social capital that originates from them, provide for individuals, groups, and organizations (Lin 2001; Lindenberg 1996; Nahapiet and Ghoshal 1998).[7]

The respondents' narratives on what factors could explain why five of the projects exceeded the initial expectations offer an abundant body of empirical evidence that strongly supports the ways in which social networks and social capital have been shown by past research to benefit innovative success. Prominent among those are achieving greater efficiency by the coordination of critical task interdependencies (Coleman 1990); facilitating access to information and speeding up the information exchange (Gargiulo and Benassi 1999); providing learning opportunities (Podolny and Page 1998; Powell, Koput, and Smith-Doerr 1996); increasing the respect, trust, and trustworthiness between actors (Bourdieu 1986; Fukuyama 1995; Lin 2001); attaining status (Baum and Oliver 1992; Podolny and Page 1998; Stark 1996); and assisting the maintenance of social norms and obligations by social sanctions (Burt 1992; Granovetter 1985).

At the same time, this pattern of organization allows the managers at Global East to avoid what Gargiulo and Benassi (1999) called the "dark sides" of social

networks and social capital. Chief among those are the likelihood of the creation of tension between formal and informal authority; the possibility that negative social networks would have a disruptive effect; and the potential for developing closed networks that could lead to inbreeding and, therefore, a loss of learning opportunities.

THE GLUE THAT HOLDS THE SYSTEM TOGETHER

This project design requires enormous flexibility from the individuals involved because of the perpetual process of reconstitution of positions and roles in the formal structure of the laboratory and on the projects. As I have shown in the previous section, a group manager, a senior engineer, or a scientist could be a manager on one project and a researcher on another that is led by his or her subordinate. It can be argued that such fluidity is likely to produce very stressful situations for the organizational members. For them to carry out their roles effectively, it means that one must often "boss" his or her boss and make oneself vulnerable by exposing one's professional self. Hence, the project structure requires people to have the ability to quickly recompose their task and role relationships, and to defy mental schemata, with their concomitant attitudes and emotions, about authority patterns in decision making.

In effect, this presents some of the quintessential challenges established in sociological theory by Robert Merton (1957)—that of social systems being able to function in a predictable way as a result of stability in social positions and the roles attached to them. Furthermore, it raises the classic question posed by Merton of the incompatibility of status-sets, and the role-sets attached to them. Less integrated social systems, he posited, will subject the individuals more often to the strain of incompatible social roles. Ultimately, such inconsistency contains the very real potential to destabilize the system and to negatively affect both individual and group performance. An "operating social structure," Merton theorized, "must somehow manage to organize these sets and sequences of statuses and roles so that an appreciable degree of social order obtains, sufficient to enable most of the people most of the time to go about their business in social life without having to improvise adjustments anew in each newly confronted situation. . . . Which social processes tend to make for disturbance or disruption of the role-set, creating conditions

of structural instability? Through which social mechanisms do the roles in the role-set become articulated so that conflict among them becomes less than it would otherwise be?" (Merton 1957: 370). With these questions in mind, I reexamined the qualitative data and my observation notes.

Such a project design, the examination of the qualitative data shows, will only be conducive to successful outcomes if supported by an organizational culture with norms, values, and expectations in place that are not contradictory to the design.[8] That type of culture does not just arise out of serendipity; much of it has to do with how a group is organized (Ancona, Bresman, and Kaeufer 2002; Hargadon and Sutton 1997; Kunda 1992). In its own turn, the adoption of this design form continually reproduces and reenergizes the very norms and behavioral expectations prescribed by the laboratory's culture. It is a culture that breeds a particular mind-set and elicits a series of behaviors and obligations that speak to the requirements imposed by the structure. At Global East, technical and managerial "stars," as well as everyone else, are expected to make time to help one another in searching for a technical solution; they are expected to share knowledge and information freely. In fact, to borrow a phrase, it is in each person's "self-enlightened interest" to do so. Although there is no immediate punishment in a formal sense, the informal sanctions of being mocked—being the target of jokes about how so and so "appropriated the knowledge from yesterday's testing," and "now we need a court order to get it"—is a potent deterrent. Neither the laboratory nor the company explicitly uses the metaphor of a family to create bonds and obligations; these are maintained through professional esteem. The mind-set is that people should care about the success of the overall laboratory and not just about the progress of their own projects. These norms and behaviors are reinforced through the procedures in place for project assignments, the formal and informal rewards and incentives systems, and the existing routines of mentoring and reproduction of human and social capital.

As noted, people at the Global East lab tend to work on multiple projects simultaneously, and the facility is small enough that everyone knows each other well. Moreover, as I showed in the previous section, the roles that lab members play on different projects do not necessarily reflect their formal positions. In fact, the lab director and the three group managers are the only

positions strictly demarcated by hierarchical lines. In general, all laboratory members are directly involved in projects. If not for the cultural values that permeate every facet of the life and the daily work in the lab, this discrepancy between formal positions within the laboratory and the project's structures could create tensions as a result of role conflict.

Assigning subordinates as project managers does not build animosities at Global East's lab, as the shared understanding is that it is in everyone's interest for a project to succeed; that everyone benefits when the lab is successful. Hence, it becomes more than acceptable for those who are recognized as the most competent in a specific area, not just by the lab's management but by their colleagues too, to lead a project. All laboratory members are expected not only to continually learn from one another but to do so willingly in order to relentlessly expand their expertise. A coveted social prize at the lab is to earn the respect of colleagues as competent and knowledgeable team players. At the same time, engineers, scientists, and technicians are continually exposed to a wide range of technical issues and the various ways in which colleagues approach them. This creates an environment of perpetual learning and the cross-fertilization of minds, as the laboratory members are also in contact with engineers and scientists from other business divisions, who may be assigned to projects at the laboratory, as well as with outside consultants. Hence, there is a perpetual flow of design and organization problems and people, one in which employees have the opportunity to develop their technical and organization knowledge and skills through project-oriented interaction, as well as to stay aware of who does what and who knows what. The knowledge gained this way is then reinvested back in the laboratory, either because it is relevant to one's next project or because this person's help or advice is requested.

Such a work environment is also reinforced by putting in place an incentive system that rewards the achievement of the overall laboratory goals, rather than the group, project, or individual goals (Rivkin and Sigelkow 2003). Everyone's individual performance is evaluated annually and is not associated with any specific project. Rather, the evaluation encompasses one's performance and professional contributions throughout the year. At the end of each project, regardless of how successful it proves to be, the team receives no rewards other than perhaps a dinner party; a formal written acknowledgement of people's

contributions, which is also disseminated through the company's intranet; and perhaps a small bonus. At this point the social recognition is what really counts, and the very prospect that one's next assignment will be a project of even greater importance and higher visibility. The importance of it becomes even greater knowing that although the director is the one who initially puts the new project teams together, members have a say on the matter as well. Thus, if a person's not-so-flattering reputation precedes him or her, it is questionable whether that person will be assigned to a high-profile project. Financial rewards and promotions, on the other hand, are tied to a person's yearly performance. Therefore, lab members cannot just be cooperative and selfless on one project; they must adopt that attitude and behavior across the board. In other words, the behavior must become more than just the *modus operandi* for a given project but the *modus vivendi* for work in general.

The practice of mentoring and the conscious reproduction of human and social capital (Higgins and Nohria 1999) are yet other mechanisms that are part of what the director refers to as "adopting a systems approach to managing the laboratory." Taking just a brief look at the personnel compositions on the six projects, and placing these assignments in some chronological order, clearly shows that the laboratory's top managers—Tom, Fred, and Olga—are engaged in a systematic development of the next generation of the organization's technical and managerial leaders while sustaining the learning environment. Young engineers, scientists, and trainees are assigned to work on projects under the supervision of other members who have a high technical and managerial reputation. This is a substantial part of the duties, not just of the managers but of all laboratory members with an established technical reputation. Each one of them works with a number of young specialists. This is how Olga describes the process:

> Their [the trainees] goal is to learn a lot of things and then concentrate on one thing. So they will need to learn broadly first, and when the time comes for a transition and when they know where they want to go [in terms of which technical area or business unit], they will concentrate on an application. But in the beginning they've got to learn broadly.

Olga's explanation shows that the process of mentoring is not left to chance but follows a well-specified logic and sequence of learning. The novices and

the freshly minted scientists are mentored so that their areas of responsibility expand with each project on which they participate until they themselves build their own reputation and become project managers. Another group manager explains the thinking behind it:

> The aim is to take bright young technical people and upscale the technical abilities of the lab and all businesses by bringing fresh blood that's got a good year of training, so when they step into their jobs they are not going, "Oh my God, what's going on?" So I had eleven people—a couple of Ph.D.'s, a couple of Masters', a bunch of engineers. . . .

Thus, new technical and organizational human and social capital are not only constantly cultivated but also put into use, and therefore recreated. For instance, after being with the organization for less than two years, and shortly after working closely with Fred, who was her mentor on *Delta*, Tonya was assigned to co-lead project *Beta* along with Olga. The project was ultimately the responsibility of Olga, who is central in both the technical and the managerial social networks. She shared the technical responsibility with Tonya, though, as a way of nurturing a member of the new crop of technical leaders.

Natalie is another example of the same practice. She was assigned as a project manager on *Epsilon* while Fred, Nick, and Olga (all technical and managerial "stars") worked on the project as team members. None of them occupied a managerial position on this project, but all three were members and, as expected by the laboratory's social norm, were available for advice, consultation, and problem solving. The ultimate responsibility for the project's success was Natalie's. An interesting observation in this case was that, despite Fred's presence on the project, the *Epsilon*'s members cited Natalie as the technically central person. Fred made certain that he was there to assist and to prevent disasters if needed, but not to overshadow Natalie, as she was establishing herself as a project manager.

The personnel assignments on *Alpha* followed the same logic. Paula had been with the organization only for a few months when the opportunity to work on *Alpha* came along. Fred recognized her technical potential and acted as her mentor throughout the duration of the project. Thus he prepared her for larger responsibilities. The same pattern is apparent on *Zeta*. Ted and Tom co-led *Zeta*. The project was Tom's idea, and yet Tom did not overshadow

Ted. Fred, who was not even officially assigned to work on *Zeta*, made himself readily available. One of *Zeta*'s members has described the situation succinctly: "Ted had the responsibility, he was in charge for this thing. But Fred had more input in the beginning." Thus, although Ted was a relatively new member of the organization, he was assigned important responsibilities, and it appears that he received all the help he needed from the person recognized throughout the organization as the one with the most technical expertise in the area. Fred did not phrase it this way, or even discuss the discrepancy between his and his colleagues' views regarding his role. Based on what I know about him, though, I think he deliberately stayed out of the project in a formal sense for he did not want to ruin the experience for Ted. At the same time, this did not create a legitimacy crisis for Ted, as this type of mentoring and learning is supported by the lab's culture. In addition, the ultimate responsibility for the project rested with Ted, so it was in his best interest to solicit and make use of all the relevant help rather than trying to accomplish everything on his own.

Such structure, culture, and processes are conducive to technological success as they reflect the nature of the R&D work and the design requirements that stem from it (Argote, McEvily, and Reagans 2003). This conclusion is backed by the data in Table 3.5. In it, I cite the managers, researchers, and technicians' answers to the question, "What advantages did you find in conducting the project in this work setting?" Their responses offer more than a clue. The answers are categorized into four major themes that emerged from the interview data. They are a supportive work atmosphere; easy access to resources, technical support, and new knowledge; a strong drive for achievement and a structure that supports it; and the creation of an organizational culture characterized by trust and respect.

In conclusion, this project design is both sustained by and, in turn, creates an organizational culture in which the social norms of cooperation, respect, and civility are upheld and reproduced. It is a culture characterized by trust and a strong drive toward superior individual, group, and organizational learning and achievement as a result of attaining status through high expert power and social capital. In addition to the processes and management practices, the creation and constant re-creation of such a work environment is facilitated by

Table 3.5

Question responses from managers, researchers, and technicians

Categories	*"What advantages did you find in conducting the project in this work setting?"*
The Work Atmosphere	• "The friendly relations between people and the physical environment" (a project researcher). • "A lot of experience and expertise are available to you, and the ability to access that at any time" (a project engineer). • "There is no deception, no back stabbing. It is a really good place to work, mostly because the way it is run." Q: "Are you serious?" A: "I swear to God!" (a researcher). • "Interacting with people. Very friendly environment. It is somewhat rare because we are in a very competitive industry. I work for [names the business unit]. I really liked working on this project because I met new people and learned a lot from them. They do not mind sharing their expertise" (a project member who works for another business unit at Global East). • "First, the people. It is a unique group of people with different backgrounds, different skills, . . . very experienced technicians. Second, the culture—as a manager you don't need to go and say it directly to the people . . . they do not need a lot of supervision. People do it [the work] for their own reason; Third, the leadership" (a project manager). • "There was a lot of friendliness [on the project]. When people just like each other . . . makes a huge difference. There was no arrogance factor" (a project and a group manager).
Access to Resources, Technical Support, and New Knowledge	• "Access to equipment, technological support, and help when needed" (a technician). • "We work on interesting and challenging projects. The boss [Tom] has a ton of ideas. Sometimes he drives us crazy because he wants to do everything. But we forgive him because he cares about the reputation of the laboratory and about us. He makes sure that we grow on the job, that we get a variety of experiences and develop a variety of competencies" (a researcher). • "The freedom to explore potential solutions, the access to required resources (people and equipment), and seeing how the proper execution and achievement of goals can benefit the organization" (a technician). • "All tools and equipment needed to perform this task were available. We did not waste any time, there was no aggravation" (a researcher).

Strong Drive for Achievement and a Structure That Supports It	• "Small group of people who are highly focused, strong managerial support, complementing personalities" (a researcher). • "It is a learning environment" (a technician). • "What is unique about this facility is the structure of the organization and the way core competencies and responsibilities are combined and how an individual can assist other individuals with what he knows" (a researcher). • "What I like most is the positive thinking and the 'whatever it takes' attitude. Personality conflicts can be devastating. Here everyone helps you and supports you. There is no 'I' in the word 'team'" (a technician). • "Just do it and beg for forgiveness later' type of attitude" (a project manager). • "My superiors are very open-minded people towards new things. They think anything can be accomplished with the right attitude. I don't think anybody here would keep anybody back" (a technician).
A Culture of Trust and Respect	• "That is one of the nicest things around here. Your opinions are listened to. Superiors consider our advice. You will find that most of the projects here are a team effort" (a technician). • "He values my opinions and suggestions" (a technician about Nick). • "I trust his opinion a lot. He knows his stuff. . . . He knows the technical and the human side of this work . . . how to manage people, you know. . . . And he has a knack for finding the best people and he pairs them up" (a technician about the laboratory director). • "Leadership is key. [Tom] is a good leader. Very smart. A quick thinker. Sometimes he does too many things at one time and gives the impression that he dropped the ball. Once you get to know him better you have all the confidence that he knows what he does. He cares . . ." (an engineer about Tom). • "No, no, she never would just tell me to do it. She would never dictate anything. . . Always discussions. She is very nice. . . . I like her" (a researcher about Olga). • "I don't think people worry about not getting credit for their work. . . . It's a very fair work environment. You feel respected here" (an engineer).

the small size of the organization, one in which people know each other really well, work on multiple projects simultaneously, and, more important, occupy different formal positions and play different roles on these projects.

In the next chapter, I explore these mechanisms in greater detail by reconstructing the histories of three of the study's projects—*Delta, Alpha,* and *Beta.*

4

ILLUSTRATIONS FROM THE FIELD: THREE CASES OF SUCCESS AND FAILURE

"The whole is greater than the sum of its parts . . . Unfortunately, not all parts fit together into a meaningful whole."
Siri N. Espy (1986: 108–9)

This chapter is devoted to the practical demonstration of the research findings reported in the previous two chapters. They suggest that the formal and informal structures on successful R&D projects are so intertwined through positions, roles, and organizational processes that attempted distinctions between the two appear rather artificial. I illustrate this point by going back to the histories of three of the study's projects—*Delta, Alpha,* and *Beta.* The outcome of each project is discussed in the light of the presence or absence of the four critical success factors—open communication across functions and levels, strong corporate support, and having the presence of technical and organizational "stars" on the project. Further, I examine these projects' formal and social network structures, as well as the social dynamics they create, in the light of the research findings presented in Chapter 3. The case of *Delta* is particularly illuminating because the project started out as a failure but was turned into a success as a result of recombining the four critical success conditions and blending the informal and formal structures. The story of the "highly successful" project *Beta* is contrasted with that of *Alpha,* a "low success" project, in order to demonstrate how their different formal-informal structural arrangements have produced different outcomes.

THE HISTORY OF PROJECT *DELTA*

Perhaps the most instructive examples of the effect of the four critical success factors, and the mechanism that weaves them together, are not the histories

of *Beta, Gamma, Epsilon,* or *Zeta,* which tell compelling success stories, but those of *Delta* and *Alpha. Delta*'s story is particularly revealing. Out of the five "high success" projects, *Delta* was the only one that had to overcome a major stumbling block in the beginning and required significant readjustments both in terms of personnel and work organization. It lasted a little over a year longer than initially planned and exceeded its allocated budget sixfold. Despite this, the project turned out to be a success. Its history speaks clearly about the importance of all four conditions to be met. What is more, it shows that when all of them are present and work together in an interrelated fashion, they create a group and organizational dynamic marked by a strong orientation toward learning and high achievement at both the individual and group levels.

The objective of project *Delta,* according to one of its managers, was "to build a machine to simulate the reaction and the decision making of a human in a particular [technical term] situation. . . . It had to help decide how the manual process works." At the same time, it was supposed to be low-cost equipment so that all branches of Global East around the world could afford it in anticipation of its wide application. It was decided to design and build the equipment outside of the United States, in a country, which I shall refer to as *Anida,* where labor is less expensive and thus the price specification could be met. As noted earlier, what makes *Delta*'s story such a strong case is that it is a project that nearly completely reversed its course. This becomes particularly evident when its two very distinct stages of development are examined carefully in terms of the presence and absence of the four conditions, and the ways in which roles and responsibilities were assigned.

Delta was a failing enterprise both technically and financially during its first stage. At the outset, the personnel assignments of roles and responsibilities differed from what they usually were on previous projects of a comparable degree of complexity. For instance, although Fred was assigned to *Delta* as a project manager from its very beginning, the range and nature of his responsibilities did not encompass the areas they normally would. Usually, he is the chief engineer and is responsible for the development and specification of all the technical parameters and test methods as well as for the engineering design. During the project's first stage his responsibilities only included the definition of the process parameters (for example, speed, type of materials, and type of force) and the broad development of testing methodology. At that

time an engineer from Anida was given the overall responsibility for designing the engineering, building the equipment, coordinating the project's activities, and managing the relationships with the manufacturers and a subcontractor who was hired to build the machine. Nick, who ordinarily assumes the latter two responsibilities and deals with all mechanical and human logistics, was not a member of the project during this stage. Such an assignment of roles and responsibilities made perfect sense given that most of the work was being conducted in Anida. Thus, giving more technical and organizational autonomy to the engineer on site was appropriate.

A little over a year after the inception of the project, the machine was built in Anida; it was not working though. As Nick reflected on the issue later, "the machine did not work . . . at all. It did not produce acceptable data." At that time, Fred flew to the site on two separate occasions and worked with the local engineer for about three weeks. The technical problems, however, continued to pile up. After his second visit, it became clear that it would not be technically viable to build this complex equipment in Anida, or from a distance. Despite the disappointing results, the divisional vice president remained convinced of the project's broad business potential and was firmly committed to see it through. It was he who made the decision to ship the equipment to America so that the work on it could continue there. Thus, about nineteen months after *Delta* had begun the machine was brought to the United States, and this marked the beginning of the second phase of the project's development.

As might be expected, this generated a lot of concern in the American facility of Global East. Lab members hoped that they would not be assigned to the project, as its troublesome reputation had become a good part of the organizational folklore. Most of the engineers and technicians assumed that it would be, as one graphically put it, "an assignment from hell." Given that lab members are routinely working on four, five, or even six projects simultaneously, not too many people had the burning desire to spend their time to "straighten up the strayed child" on top of their regular workload. Moreover, there was the added concern that members' next project assignment is based, to a significant degree, on how well they prove themselves on their previous one. Because *Delta* was a "high visibility" endeavor, few lab members wished to be associated with a project that was losing ground, "sinking for sure"; furthermore, the divisional vice president took a personal interest in its successful outcome, which only

increased the pressure on everyone involved. It was at this point that Nick was assigned to the project and given his normal responsibilities—to coordinate all the logistics, to identify a suitable subcontractor, and to arrange for the necessary tests and equipment. Fred, as well, stepped into his "usual role" as a chief engineer and assumed all the responsibilities of the engineering design.

Both Nick and Fred shared a number of concerns regarding the financial and technical prudence of continuing the project by working on the machine shipped from Anida instead of building a new one from scratch at the lab. This decision, however, had been made for them by the laboratory director and the vice president, so they proceeded. In their judgment, the equipment had to be at least rebuilt, which in many ways is more challenging and demanding "than starting from scratch." Despite the difficulties, the work on the project started to pick up and results were beginning to show. This is what Fred had to say about it:

> We had to rebuild it because there was a lot of concern about the quality of the machine, because it was being made in Anida—whether it would be pieced together with rubber bands and paper clips, . . . or whether it would be a really quality machine. Generally machine equipment that comes from that part of the world is not very good. And it did come that way, so it needed a lot of mechanical rework when it got here. . . . It was a nightmare to maintain. I mean . . . every week there was something that would break on it . . . and Nick had to maintain it.

Nick was convinced that had Fred and he started working on the project from scratch, it would have taken less time and financial resources and would have saved a lot of aggravation for everyone. In the following statement he explains the importance of the combination of technical and organizational competencies on the project and the value of good and direct communication:

> Well, we talk to each other, not past each other. What I am saying is that it takes very little for each of us to understand what the issue is and then we know what to do. It is mostly Fred who makes the technical decisions, and I make sure that he's got what he needs and whenever he needs it. . . . I deal with the manufacturers, suppliers, scheduling . . . Fred concentrates on the design. . . . You can't really do this with overseas projects . . . So, there was a lot of guesswork, which makes it harder . . . and *very* [emphasis his] frustrating. . . . It takes so much more time.

Twenty-one months after the equipment was brought to the laboratory, the machine was "up and running," producing simulation data, and "refuting," as one of the project managers had phrased it, the people "who kind of gave up on it." Ultimately, *Delta* became a financial success and helped to uphold the reputation of the R&D laboratory and the company. As one of the company's vice presidents stated, after the project was completed several of Global East's divisions worldwide became "aware of the great people in this facility who solved this 'impossible problem.'" But can this spectacular turn-around be solely attributed to the technical wizardry and organizational skills of Fred and Nick? Is it plausible that the two of them could have achieved it alone? The reactions of the project members confirm what I had heard so many times in the interviews—"it is a team project," "we worked as a group," "we stuck together." The interesting question, though, is What was the glue that made them "stick together," given that working on this project "was a nightmare for everyone 24/7"? At more than one time, people were ready to give up, especially when the decision was made *not* to start working on the equipment from scratch, but to fix and rebuild the "rubber-banded temperamental piece of metal."

During *Delta*'s first stage, one technician had already left the project, as he could no longer deal with the stress created by the unpredictability of the performance of the machine and the never-ending changes in the specification of the technical parameters as a result of "constant communication gaps on the route between [Global East's R&D lab] and Anida." Nick was not even a member on this project, let alone assigned any managerial duties; it was not entirely Fred's "game" either, as it was the responsibility of the engineer from Anida. The work on the project was not problem-free during the second, post-Anida stage either, when both Nick and Fred were the managers. All hurdles in terms of planning, scheduling, coordination, time, and cost never really ceased. Everyone was frustrated and exhausted. One of the managers explains,

> It took a lot of time; it tied up a lot of resources to get it [the machine] running.... Scheduling was a nightmare; it took a long time to get this machine to demonstrate what it could do.... [We] had to hire another person to fix it, so that we can have the people available to work on other projects. We are a small facility. People work on several projects simultaneously.... We always needed more people to keep all the other projects running along with this one.

Despite all obstacles, the project triumphed. Could this success and perseverance be attributed to the reputation and respect that Fred and Nick commanded across hierarchical levels and throughout the organization? In fact, many of my respondents referred to both of them as "leaders" as opposed to "managers" when they discussed the work on this project. As a young engineer put it, "You don't manage such a mess; a manager can't get you out of it. The only way to save the day is to have a leader—someone you trust to lead the way out of it." The words of a researcher on the project speak for themselves:

> They [Fred and Nick] are very positive. They give you confidence. . . . Both of them. You go to them and give them the bad news—something broke . . . again . . . [laughs]. You don't expect what's going to happen is they'll kill the messenger. You don't expect to be accused. They listen, then we talk, we brainstorm, make a decision and we go try it. . . . until we solved it . . . O-o-h, yeah . . . we yelled at times and we talked a lot. I even wanted to kick it [the machine] . . . to vent out. Then we laughed and went back to work. . . . That's life in a technical world.

The same sentiment is captured particularly well by a technician on *Delta:*

> We were kind of losing hope with this machine. . . . It was a never-ending ordeal. . . . Every day there was a new problem with it. You fix one thing, another breaks. . . . You solve one problem another pops in. . . . It was kind of disheartening. . . . We were very frustrated. Nick told us he would hire outside contractors to work on it and to give us a break. . . . We were really losing it with this project. . . . Break sounded good, but we could not let these guys [Fred and Nick] down. They were frustrated too. I just believed that if they wanted us to try something they had a reason to. . . . We could not let them down. . . . They would stick up for us any day. . . . I am glad we all stayed on. We proved ourselves. . . . We did not let them down.

The essence of the laboratory's culture is captured in these statements— that is, "respect begets respect." Despite the frustration that was building up daily and the choice to step out of the work on the project, no one took the chance. Just the opposite—four more scientists and technicians were added to the project's staff after the machine was shipped from Anida, and no one left prior to the project's completion. The team did, indeed, stick together. But what afforded them the opportunity to "prove themselves" given the

one year of continuous technical failures? Could the success on this project be explained by the factors I have discussed so far—technical and organizational expertise, open and direct communication, and mutual respect? The accounts of the project members clearly suggest that this is not the case. These are all necessary, but not sufficient explanations. Had the project not received the extraordinary support of the divisional vice president who kept its budget up, no one would have had the chance to prove their technical skills, to "stick together" with the "leaders" they respected, or to taste the success of the outcome of their work more than three years later; *Delta* would have been cancelled. As Nick has put it, in a business environment "you constantly worry about determining the priority projects and the priority of our resources. . . . it is very difficult to fix that." Any investment decision mistake leads to financial losses for the company, losses of reputation and referral business, and losses in salaries and jobs.

Someone from the R&D laboratory with a solid technical and organizational reputation had to assure the vice president of the importance of this project, its wide technical ramifications, and the warranted nature of the risk to be taken by continuing to support it. Someone in a high, formal position of power had to assume the responsibility and to vouch for keeping the project on "life support." If it were not for Tom's political skills and technical accomplishments, the divisional vice president would have most likely cancelled *Delta* after the first year of continuous mishaps regardless of his initial interest. As Tom himself succinctly phrased it on several occasions, "Success is also political, not just technical." And what gave credence to Tom's assurance that the project could work was that he had in the laboratory these two technical and organizational reputation cards—Fred and Nick—and he could play those out. Thus, people with widely recognized competencies are viewed as contributing to the successful outcome of a project even before a project has begun, even before they are formally assigned any roles and responsibilities.

Delta's history shows that it is not enough to have the right ingredients to achieve success; it is the way in which they are put to work, the way in which they are forced to interact; it is the glue that keeps them together—the culture that places a high value on support and achievement and the social capital that emanates from members' centrality in the two critical laboratory advice

networks. In this manner, the social capital is not only recognized but accessed and mobilized to produce results. If it were not for the well-deserved reputation that Tom, Fred, and Nick enjoyed from the corporate management, the budget of the project would not have been constantly replenished. If it were not for the respect that these individuals enjoyed from the people in the laboratory, it would not have succeeded either. Working on such a complicated, difficult, and prolonged project tests everyone's patience and loyalties. Emotions run high in such circumstances. If you do not trust your leaders—their technical, organizational, and human competencies—sustaining such an effort would be inconceivable. These are the times when open communication channels pay off because they "cut off the work of the rumor mill" and provide an outlet for the aggravation, in addition to serving their basic function of keeping everyone well-informed and up-to-date on the work and progress of the project.

When the project ended, its technical outcome was a new piece of equipment and an innovative technological process. As a result of it, a new business group was created that tests different products with the equipment designed as a result of *Delta*'s completion. Here is how Tom, the laboratory director, describes *Delta*'s outcome:

> What came out of it was a range of capacities for product development for a range of products. Along with this, now Anida has a capability for testing [the product], which they would not have had without participating in this project. Essentially, what we are trying to do is leverage the capability of . . . site-specific development . . . integrate it, leverage it to develop other sites.

Although it had never been used for the purpose it was initially intended for, the newly created equipment found a wide application in another business division and proved to be instrumental in new-product development in various locations of Global East. The technical outcome of the project continues to generate revenue for the company.

THE HISTORY OF PROJECT *ALPHA*

Alpha, a "low success" project, started with a clear technical objective—to modify an existing product and to drastically improve its performance characteristics such as speed, size, and precision of operation. Five R&D members

worked on it: Fred, who was the project manager; a young scientist who was also a trainee; and three experienced technicians. Time constraints were of serious managerial concern, as the identified market opportunity was beginning to narrow. For projects that represent more routine technologies, such as *Alpha*, staying within the time frame always is a matter of business urgency. The lab director makes this point by arguing that different "kinds of pressures" are exerted on this type of project, ones that are

> more related to effectiveness, timely delivery of results, strict deadlines, kind of the sooner the better. The sooner you can do it, the sooner the customer can do something with it. It is a strict deadline because if you don't do it your window of opportunity is gone.

Thus, timeliness of a project's completion is directly linked to its ability to have a market impact. *Alpha* resulted in the creation of a new product, but it was evaluated as a "low success" because it only met its technical expectations, it did not exceed them. It did stay within its projected time and budgetary constraints, but had no market impact, let alone one of the magnitude of *Gamma*, *Beta*, or *Epsilon*. What clearly differentiates *Alpha* from a "truly failing" project is that it has not lost money, but it has not made a profit either. Moreover, there was very little expectation on the part of management that this would change. No corporate officer acted as a champion for this project and no one recognized by the laboratory members as a "managerial star" was assigned to work on it. Because the project was considered to be "a routine technology" from a technical point of view it was assigned only one manager—Fred, a technical "star."

The idea for this project came from a customer who was apprehensive about being too dependent on a single European supplier, with whom the customer had had a relationship of several decades. There was no commitment, though, on the part of the customer. As a result of it, top management looked at this project as an opportunity for the company to demonstrate technical capability as opposed to an opportunity "to shake the market." It is, perhaps, the reason why no one from the upper echelon of the corporate administration became a champion for it. Fred, on the other hand, saw the project as a chance to demonstrate technically that this problem could be solved in the lab and that it could have real business potential. He was determined to modify

the existing product in a way that would introduce drastic improvements in its performance.

In retrospect, the absence of strong corporate support and a person with organization skills on the project affected the way in which *Alpha* was managed and developed. In a technical environment in which "a good portion of the work is a matter of trial and error," the essence of corporate support is that "when you have a sufficient amount of time and money, you can run all the technical tests that you need and be confident in the results." The importance of corporate commitment for the project's financial viability is also reflected in the following statement offered by a project manager: "On some projects the money is not an issue. But sometimes, when you do not have full corporate support, money becomes an issue."

Alpha had to adhere strictly to the timetable and was not granted the additional latitude requested to conduct more detailed experimentation, for an extension of the duration of the project would have meant additional funding, and no high-ranking corporate officer took this chance. Fred attributes *Alpha*'s outcome to the insufficient time the team members were given to bring the project to a close—"about nine months and not a day extra." This is Fred's narrative, in which he expresses his lament for the lack of time. As usual, his account is a complex one and touches upon several interrelated factors:

> On this project here [points to Alpha], money was not an issue. Time was . . .
> Time and the likelihood of success. There was concern about how much effort will be required to satisfy the customer. . . . Technically, I would have been happier . . .
> if we had done a greater amount of testing to evaluate that the process repeats better, that the results were consistent. It would have required more time. There is always a compromise that way. People who request that type of projects don't want to spend the money, or the time, and people who work on them, especially when you are . . . when you are my type of mentality . . . you want to make sure that the answers you give are absolutely correct. . . .

Although Fred never phrased it so, granting more time to work on the project, from the company's perspective, would have clearly meant additional financial resources to cover the extra labor costs. The first tier of decision makers, obviously, were not of the opinion that the allocation of more capital

investment in this project was prudent. Fred, on his part, was convinced that a few extra months were crucial:

> I knew we could do it. . . . Tom knew we had the expertise to do it. It was not such a difficult issue. It was a matter of running many trials. . . . We were in a very good position to solve this. . . . We knew how to do it. . . . We had all the necessary expertise. We could have had a product now X times faster and X times more precise, but we did not have the time to run all the tests. . . . I guess, they [corporate management] just could not see it that way.

He links closely the time and political support factors. For Fred the lack of corporate understanding translated into an action, or *in*action rather, that resulted in not having enough time and, ultimately, to the fact that the true potential of this project was never fully developed. One might speculate though that had there been an organizationally central person on *Alpha*, who would have taken care of all the logistics for Fred and kept him within the time frame, he perhaps might have had the additional few months to do more testing and be satisfied with the results. Neither Nick nor anyone who is central in the organizational advice network had been assigned to this project. Given Fred's dislike of "feeling like a manager," and his "engineering mentality," which does not let him rest until he exhausts checking all possibilities, it is conceivable to think that someone like Nick, whose passion is creating perfect organizational logistics and "pushing the work," would have been much needed on the project. Fred himself acknowledged that his perfectionist nature will keep him testing until someone stops him. A lab member described him as a true follower of Albert Einstein's motto: "No amount of experimentation can ever prove me right; a single experiment can prove me wrong." In the absence of someone like Nick to keep the timeline straight and build some pressure, Fred "will always be willing to do more testing and to try new things."[1]

Fred's dislike of managerial duties is also reflected in his devotion to mentoring and to the fostering of the next generation of "technical stars" in the laboratory. He finds a lot of satisfaction in mentoring, and as one of the laboratory members pointed out in his narrative earlier, Fred did indeed leave the previous company he worked for when he was promoted to a managerial position. To him, that meant less engineering design work, very little mentoring,

if at all, and a lot of paperwork. During his interviewing process at Global East, Fred was not shy about expressing this sentiment. He accepted the position they offered him only after Tom and he agreed that even if Fred's title is "project manager," he will "be doing the work" and he "will not be breathing down people's necks." In fact, mentoring is what Fred considers to be one of his responsibilities in the organization. His experience on this project was no different. By his own account, he found much satisfaction in mentoring Paula, a young engineer, who had joined the organization just a few months prior to the beginning of *Alpha*. When I asked him what "the top three things that he valued most about his experience as a project manager on *Alpha*" were, this is what he had to say:

> Well, certainly one of them was the satisfaction . . . of mentoring Paula on a project from scratch. Because it's . . . I mean . . . the satisfaction of working with someone who takes a little bit of direction and runs with it and does a good job. She is very intuitive and very aware . . . and this is only a couple of months after she started here. So, it was very satisfying to work with someone who was enthusiastic, and intelligent, and insightful into what we were doing and could work without a whole lot of direction. And then, there is the satisfaction of seeing this [names the technical issue] being done inhouse, of showing that it is possible. . . .

The statement clearly reveals Fred's passion for designing new technology and passing this passion on to bright young scientists and engineers. *Alpha*, after all, met its technical expectations. Its "low success" outcome was attributed to its inability to develop a product superior to the one that the customer's supplier was offering. The combination of a lack of strong corporate support, and the absence on the project of someone who could take care of the organizational logistics, coupled with Fred's perfectionism, could perhaps constitute a plausible explanation of why there was not enough time to run all the necessary tests and to be as confident as possible in the performance of the machine. On the other hand, given this project's technical strategy, which is perhaps more accurately described as "exploitation of old certainties" as opposed to "an exploration of new possibilities" (March 1991: 71), it could be the case that assigning a technical star on this project was counterproductive in the first place.

THE HISTORY OF PROJECT *BETA*

The objectives of *Beta* were to develop a new, radically different product and to make it commercially successful. Olga and Tonya co-led *Beta*. While Olga had the major responsibility for the project, both technically and organizationally, Tonya was instrumental in terms of executing the technical side of it. Although not recognized by her colleagues as a "technical star," she was very prominent in the lab's instrumental social network, obviously a reflection of her strong technical background. Tonya was assigned to this project because of her substantial knowledge and experience in the particular technical area relevant to *Beta*. In fact, she was hired at the R&D laboratory less than two years prior to the beginning of *Beta* because of her strong and highly specialized skills in two very specific technical areas. One of these was underutilized in the laboratory, and management wanted to develop projects in this direction; the other one was relevant to *Beta*.

Beta was a spectacular success. This is how Tonya describes its outcome:

> Technical success is the major thing. There were a lot of technical challenges, but we solved them. And our customers have not seen anything like it. This is what made this project so successful. Technically, it is so superior that a new business was created. . . .

The idea for this project came from the corporate R&D and the business division that the lab is servicing. As the lab director explained, "the goal was to bring the technical image of Global East in [names technical area] and bring it up in technological prestige." It was a high-visibility project from the very beginning. A group manager explains the rationale for it in the following manner:

> It was technically challenging but with a lot of potential. They [the two divisional vice presidents] wanted to have a superb product and a very targeted product launch. They wanted to be able to say, "This is a [name] product," to give it to a sales guy, the sales guy goes out to five customers, and they wanted at least four of them to be a success. So they wanted the product launch not to be a shotgun, just taking it everywhere and failing half of the time. They wanted it to be a success. One of the worst things that can happen to a new product is if you take it out and it fails a lot. The business, the market areas can be close enough that this customer

says to that customer over a buyer, "I tried that product, it was terrible." But that customer has a different application from the other customer, so you never know whether or not this product could be successful among customers or have a negative connotation about a product. So he [the vice president] really didn't want that to happen.

There was some reluctance on the part of the lab director to take on the work because of its high visibility and because, as he put it, "we were not up to that point much of a [technical term] center," and the lab had not developed a distinct expertise in this specific technical domain. At the same time the stakes were high; not one but two of the company's vice presidents had a grand ambition for this project; they wanted to change the way the business worked. In short, the project had to succeed. The technical and organizational challenges for both the business units and the R&D facility were enormous because the technical problem itself had not been tackled before. As Olga explains,

> They [the business units and the vice presidents] were putting a lot of eggs in the basket here. They wanted to set up a whole manufacturing center, a new business. They were really trying to make a departure from standard [technical term] practice. They wanted something that is high-tech product, high-tech image, and a high-tech manufacturing facility. That is a big amount of investment.... There were a lot of people looking at us....

Thus, there was a lot of pressure even before the project had begun. The vice presidents, Tom, the group managers, and managers from sales had gone through rounds of discussions and brainstorming. One of the group managers, a self-described technical geek with a distinct taste for challenging assignments, had rooted for the project ever since the idea surfaced. She had unwavering confidence in the ability of the lab to take on such an endeavor and spent a lot of her human, social, and political capital to convince the rest of the decision makers of it: "And for us to be at that interface where it is going to be launched successfully was great." *Beta* had officially started. The techies in the lab were thrilled. A researcher said, "This was one large project that came in our way, and I was very pleased to get a project like this." Yet another one stated, "A project like this . . . fascinating! You get to do really cre-

ative stuff. . . ." A third one described the rewards of having the opportunity to work on a project of such importance in the following manner:

> There is the satisfaction of proving that we could do this inhouse. . . . And, I don't know . . . there is also the satisfaction in feeling that other people, like my boss, see you involved. . . . I don't know how you would put that.

Unlike with *Delta, Beta's* launch was marked by excitement, creative tension and apprehension, high hopes, and a strong desire to succeed. The work on the project generated a lot of positive energy, which reverberated throughout the lab; many people wanted to know what was going on with it, what had been tried, what had worked, and what had not. It took a little over a year of intense work to finish this project. Despite the very tight schedule it was completed on time "without all of us losing our hair," as a technician described it. When it ended, the outcome of the project was described as threefold: a new, radically different product came out of it, and a new measurement methodology and a training component were developed.

Here is Olga's account of what qualifies this project's outcome as a "high success." The success elements—technical, managerial, and market aspects, as well as support from corporate management—are all intertwined in her narrative, perhaps a reflection of her position in the formal organization structure as a group manager and her position in the social network structure—that of centrality in both the technical- and organizational-advice networks. She states,

> The most important question is whether we achieved our technical goal, what we said we would do when we sat down with the people from the business division and sales. . . . Did we do it when we said we would, and did we have to put a lot of extra man hours into the project? . . . Did it cost us more than what we expected? We had flexibility on that one though. The vice president wanted it to be a success. . . . It definitely exceeded the expectations. The sales are going up very quickly.

In fact, the technical outcome of the project was "a product so superior to what was on the market until then" and so technically sophisticated that the corporate management decided not to take any chances and to train the people who were going to promote and sell it. As a researcher explains, "the

vice president made the decision that nobody could sell his product without being trained." Olga elaborates further on the necessity of such training and its direct relevance to the business goals of the company:

> There are salespeople, and the way they get business is to take people out golfing and schmoozing and dinners . . . and so on. That's just a reality. There are technical sellers, and there are personal sellers. It [names the type of business] is not an art, it is a science. So you need technically knowledgeable people to promote your product, not just schmoozing charmers.

Thus, a third dimension was added to the project—the training of the sales-force.

Beta's outcome was judged by R&D and Global East's managers to be a radical innovation, both technically and organizationally; "a pretty major de-parture from existing practices technically. It is a radical departure from the common way that a product is launched too." This evaluation is not surprising given that a new business was created as a result of the product developed by *Beta*'s team. Olga's view is supported by "the facts": "[the vice president] has a totally new manufacturing facility, a total departure from the conventional manufacturing of [technical term] products." The project exceeded not just the technical expectations but the financial ones as well. "The sales are going very quickly," Olga continues, "It's max-ed up. It doesn't have enough capacity to do everything because it has taken off so quickly."

The work on the project followed the "normal frenzy." Seven lab members were on the team. Consistent with the lab practice, each of them worked on other projects simultaneously as well. As a manager describes it, "The number of projects that we actually work on in the lab is huge, and we are busting our ass constantly. Tonya, for instance, probably has five or six projects that she keeps going." To complicate matters further, those lab members who have de-veloped a particular technical expertise are also assigned trainees to supervise and teach while working on a project. The flow of manpower in the labora-tory is dynamic. In the words of a group manager, "It is a little bit scary to me sometimes."

How did the people handle it? The narratives of the lab director, and *Beta*'s project managers and members, unmistakably point to the four critical success factors. Perhaps it all begins with the unwavering support from the two vice

presidents who were determined to see this project succeed. Olga expresses its importance rather well:

> That's why I am saying that it was a high-profile project, because there were these two very big guns who had a vested interest to succeed and who did not have patience. They did not have the opportunity to let it out to fail.

As was the case on the other successful projects, when corporate support was firm, the budget was not an issue, and no resources were spared. In fact, *Beta* did not even have a budget in the traditional sense of the term; allocations were made on the "whatever it takes" principle. Such a high degree of corporate support also translated into complete access to the decision makers at all times. Throughout the entire duration of the project, and the training component as well, the communication channels at all levels were kept open and working. Tom, Olga, and Tonya had unrestricted access to the vice presidents. The communication flow between the project members themselves and the project managers followed the same pattern.

The project developed in two stages. The first—"the fuzzy part"—was the one during which all technical possibilities, and ways of addressing them, were discussed, and the technical parameters and approaches were evaluated by all the people with the relevant technical and organizational skills and knowledge. The second phase consisted of the work when answers to these questions had more or less emerged to the satisfaction of all parties involved and the job became more a matter of the execution of the plan. This is Olga's recollection of the pattern of information exchange and communication that took place during the time when the team was setting the technical objectives:

> Tom and I would sit down, and we would put together what we thought was reasonable, what we thought we can do, how to do it, and we would review it. Then we [would] discuss it with the business people. They made suggestions, and we had discussions of how many people we needed and how to organize it. This was almost daily.

Afterward, the two project managers would meet and continue the discussion as many times during the day and the week as needed. If they required to clarify things with the business division, they had full access to the vice presidents and all people there. Olga describes the process in the following

way: "Tonya and I spend a lot of time in discussion with him [the business vice president]. I enjoy working with him. There was a lot of friendliness. . . . There was no arrogance factor." Again, at the project level it was no different:

> Talking to people on the project was easy too. We all wanted the same thing—this project to succeed. We all need each other's input. No one can succeed on his own. So it was just that people like each other . . . it makes a huge difference.

These free-communication channels reinforced in the project members the idea of the importance of the work they were doing and reminded everyone of the enormous responsibility they shouldered. This, however, did not have a stifling effect, as it also provided a venue for open discussions at any level. Hence, it created a working environment in which project members were clear not only on *what* had been decided that they had to do, but also *why* a particular technical approach was discussed, what advantages and disadvantages each member saw in applying it, and so forth. This leadership style, which incorporates not just open communication but a two-way exchange of ideas, gave the project members a sense of ownership in addition to the responsibility. Each felt that their contribution would be recognized; each wanted the project to succeed, as they saw themselves succeeding if the project did. Here is how Tonya describes the way in which Olga and she co-managed *Beta:* "We always had to agree on things. She never would just tell me to do something. She would never dictate anything; we always discussed things." This was further supported by a fluid and flexible project structure, the shape of which facilitated the tasks that needed to be completed. When asked to describe the organizational structure of the project, the senior project manager simply said, "Nonexistent."

During the second phase of project development, the communication patterns and leadership style changed to accommodate the different nature of the work. Tonya describes the shift in the following way:

> At the beginning we [she and Olga] talked about the technical issues and what's to be done every day. It was a huge project, and we wanted to get it right. I wanted to make sure that she was okay with it, so I would tell her what I was doing everyday. It is like an open discussion, and just everyday we discuss what needs to be done. Later on . . . maybe once a week. And it was monthly when it became all streamlined and the only thing that needed to be done was just click on it and

then do it. . . . So at the end it was just making sure we were meeting the deadlines, that was all. We didn't have that much to discuss technically.

This ability to change supervision gears so easily is a result of the deliberate strategy in the laboratory to promote self-reliance, "to give people a sense of ownership" and "not to treat them like incompetent five-year olds." Such an approach, Tom says, is good for the company, as it allows "young scientists to take on more responsibility while not solely held accountable." Olga describes the sources of her satisfaction from the work on the project and the level of her involvement during each stage in the following manner:

> Working with people and actually seeing them getting involved, getting a clue . . . was very satisfying. I was very involved in the first, say, six months, participating completely; making decisions, discussing options, just everywhere on the map. . . . And then I started to phase out because it was not necessary for me to be there. Once I was sure that everything was okay then I decided to leave it up to them.

The project members, regardless of whether researchers or technicians, paint the same picture of autonomy coupled with open communication and discussion pattern, one that involved all the people with the relevant knowledge, expertise, and decision-making capacities. After things had become more or less clear, a shift in the responsibilities on the project also took place. For instance, during this stage, it was Tonya who would address all the technical issues: "I will decide technically what I need to do. I ran the project from then on; all the technical decisions, I made. All the managerial decisions, she [Olga] made." Olga remained responsible for keeping it together, and thus she continued to make all the financial decisions and the personnel assignments. The level of supervision lessened substantially, and the project members were left to "follow the methodology" that was agreed upon and to "do the analysis."[2]

Throughout the duration of the project both managers and project members could rely on having access to each other, to the lab director, and to the two vice presidents. No single person was given the responsibility to coordinate the activities between the R&D lab and the business unit; it was everyone's duty to convey the relevant information. As a project member describes it, "We talked to each other directly when needed. There were no hierarchical hurdles. It would not have been successful otherwise."

Without exception everyone involved pointed out how critical it was to have the people with the right set of technical knowledge and skills, as well as the ability to put those together and to make the "project sail." The people with the strongest reputation in these areas, both within the R&D laboratory and in the company, were put on this project. This was true in terms of both technical and organizational skills. An engineer states,

> You can't do one without the other. . . . If, for example Tonya was not the person working on that project it might not have done so well, because she had all the necessary knowledge. . . . She has unique technical expertise without which we could not do it. If it wasn't for Olga's experience and skills—I mean technical and to deal with people—we could have easily lost it. . . . In retrospect, I wish we had more realistic time estimates for completion but, overall, it worked as smoothly as it could have. No tension between people, there was a lot of trust.

A project like this one could have been a nightmare. There was so much pressure in terms of both time and results. Two vice presidents were watching it closely, "coming to the lab, asking questions . . ." *Beta's* story can best be concluded with Olga's assessment of what accounts for the project's extraordinary success. She wraps it up as follows:

> We just happen to have a really good team. You will probably say a very small load, but a critical load. . . . Knowledgeable, willing to learn, not arrogant . . . very competent people. I honestly think that the key was a lot of cooperation between the people here [in the laboratory] and the people in business. It was 100 percent direct communication across the board. If they had been hands off and not participating, it would have been not very successful. You need their input. . . . It was high profile because the vice presidents would not let it fail. Failure was not an option. . . . There were a lot of people looking at us. And, when you have a pool of talent—engineers, researchers, and managers—and you put all these [things] together . . . it usually works. . . .

Hence, without being prompted or led in any way, Olga marked all four milestones for this project's success: (1) direct communication at all levels—between the members on the project, between the project members and the project managers, and between the project managers and the two relevant business units; (2) a superior technical expertise; (3) a pool of organizational talent that could put all the elements together; and (4) the unreserved support

UNIVERSITY OF STRATHCLYDE

ISBN	Qty	Sales Order		
9780804755702	1	F 9070496	1	
Customer P/O No		Cust P/O List		
87137/VLI93229LC		38.50 GBP		

Fund: ABK404

Title: Secret of success : the double helix of formal and

Format: Cloth/HB

Author: Rizova, Polly S.

Publisher: Stanford Business Books,

Volume:

Edition: **Year:** c2007.

Order Specific Instructions

Ship To: 28995001 F
UNIVERSITY OF STRATHCLYDE
CURRAN BUILDING
101 ST. JAMES' ROAD
GLASGOW
STRATHCLYDE
G4 0NS

Bill To: 28995000
UNIVERSITY OF STRATHCLYDE
CURRAN BUILDING
101 ST JAMES ROAD
GLASGOW
STRATHCLYDE
G4 0NS

Routing

Sorting
Y04G06X
Covering — BXAXX
Despatch

from the "two very big guns" who "could not afford to let the project fail." Yet again, her narrative suggests that if one of these factors is out of place, the winning equation is going to be disturbed; it is only when all of them act together that the energy is generated to mobilize the resources necessary for a project's success. And it did succeed because all those with superior technical expertise were in a position to make decisions; all those who were part of the "pool of organizational talent" were the occupants of formal positions that would give them both the access and the authority to use their talents to the full.

In the next, and the last, chapter of the book I discuss the theoretical and practical implications of the lessons that can be drawn from the study of the six cases at Global East.

THE LESSONS OF SUCCESS

"True innovation is complex and tumultuous—full of spurts,
frustrations and sudden insights. . . . Somewhat orderly chaos."
James B. Quinn (2000: 22–24)

A SUMMARY OF THE LESSONS

In this book, I set myself the goal of examining what factors, and in which combinations, could best explain success on R&D projects. The main focus of my investigation was the dynamics created by the combined effect of the formal and social network structures and the ways in which they shape the outcomes of innovation. In particular, I sought to explore how the acquisition, sharing, and utilization of knowledge in R&D environments is constrained or enabled through positions of centrality and allocation of roles and responsibilities in both structures. The main conclusions from the study can be summarized by three central findings.

The first, as reported in Chapter 2, concerns four critical success factors that emerged from the analysis of the interview data. One element that is crucial to the success of R&D projects is a heavy reliance on open and unrestricted patterns of communication, coupled with a low degree of formal reporting. Another condition that turned out to be critical is having on the project individuals who are highly respected across the laboratory for their exceptional technical skills and competence. Similarly, it is also vital to involve in the project those who are highly respected for their organizational expertise and experience. A fourth and final key factor is strong and sustained support for the project from the company's corporate management. What is more, the nar-

ratives of the laboratory members gave particular emphasis to the interactive nature of the effect of the four conditions; namely, that they are not likely to produce successful outcomes on their own but only when put together in an interactive relationship, each one being a prerequisite for the others. Although existing empirical studies on innovation and project management have identified that open communication patterns, strong support from corporate management, and "stars" being assigned to projects are critical to technical success (Brown and Eisenhardt 1995), it is the nature of their relationship that adds new knowledge to the literature on innovation and project management. Furthermore, the findings from prior research on their positive effects have not been consistent. In fact, as I have shown in Chapter 2, at times they are even contradictory. Thus, to test for both the combination of the four factors and their interactive effect, I employed another analytical methodology—Ragin's Qualitative Comparative Analysis (QCA) (2000, 1994, 1987). The results from the QCA lent support to the claim that the "causal combination" of these four factors will be "sufficient more often than not" to produce a "highly successful" project outcome.[1]

I then explored the question of *how* the four conditions were placed into an interactive relationship. From the literature on innovation and project management we know that simply having assigned to a project individuals with high human capital—technical and organizational "stars"—does not guarantee that they will perform and that a favorable outcome will be achieved. Rather, as the social network literature would predict, it is the relations between the people with the appropriate knowledge and competence that matters (Saxenian 1996, 1988). Further and more detailed analysis of both the qualitative and the social networks data, which I discussed in Chapter 3, confirmed that the presence of the four factors will not, on their own, lead to a project's success. Not only must each of the four conditions be met, but they must be put together in a way such that they reinforce each other. The question then became, *How* does this come about? Mapping the results from the social networks data onto the findings from the qualitative analysis reveals that success on R&D projects is more likely when the formal and the informal structures are interwoven in a specific way. At the laboratory, this is achieved through a unique project design; by assigning positions, roles, and responsibilities on

projects in accordance with an individual's human capital and centrality in the two work-related-advice social networks—*technical* and *organizational*—that I specifically created to reflect the nature of the knowledge and information exchanges, and the work processes, in an R&D setting (Rizova 2002). In other words, the positions of centrality in the *actual* advice relationships are converted into positions of centrality in the *prescribed* ones. The uncovering of this mechanism is the second finding from the research. This finding is surprising given that traditionally the social networks are viewed as competing with formal structures, work processes, and an organization's culture (Cross, Borgatti, and Parker 2002).

Moreover, a closer look at the practices at the laboratory for project assignments revealed that the roles that the engineers, scientists, and technicians play on different projects do not necessarily reflect their formal positions in the organization's hierarchy. In consequence, a senior engineer or a scientist may be managing one project and at the same time be a researcher on another that is managed by a subordinate of his or hers. I found this result to be both intriguing and puzzling, as it raises the classic question posed by Robert Merton (1957) of incompatibility of status-sets, and the role-sets attached to them, and the potentially destabilizing effect of status and role inconsistency on individual and group performance. Therefore, I needed to find an answer to the questions of whether and how the members' experiences of role conflict, as a result of the constant shifts in positions and roles, had an effect on the outcomes of the projects and on the social climate in the laboratory. This led to the third finding.

Further investigation of the interview data revealed that this type of project design will be conducive to successful results if it is supported by an organizational culture and institutional processes that generate norms, values, and expectations that are complementary to the design (Argote, McEvily, and Reagans 2003). In its own turn then, the adoption of such a design form continually reproduces and reenergizes the very norms and behavioral expectations prescribed by the laboratory's culture. It is a culture that breeds a particular mind-set and elicits corresponding behaviors. At Global East, technical and managerial leaders, as well as everyone else, are expected to help one another in the pursuit of technical solutions; they are expected to continually learn

and to share knowledge and information freely and are recognized for it. The mind-set is that everyone should care about the overall success of the laboratory and not just about their own projects. These norms and behaviors are reinforced through the existing routines of mentoring and by the procedures in place for project assignments, as well as through the formal and informal systems for rewards, incentives, and sanctions. It is a culture that promotes its code not as the *modus operandi* for a given project, but the *modus vivendi* for work in general.

THE LIMITATIONS TO THE LESSONS

Before I turn to a discussion of the implications of the findings for organization theory, and the literature on innovation and project management, I would acknowledge some caveats of this study. The immediate applicability of these findings to R&D organizations and projects is limited in several ways. The first and most obvious one is related to its methodology. This research represents an inductive case-oriented comparative approach, and as such, "the goal of appreciating complexity is given precedence over the goal of achieving generality" (Ragin 1987: 54). Furthermore, my choice of relying on this research strategy was motivated by the need that I identified, which is also acknowledged in the project management and social networks literature, to raise fresh questions and to carve a space for a discourse about new theoretical and practical possibilities in the study of technology (Powell and Grodal 2005; Nebus 2006). Consequently, immediate generalization of the study results must be treated with caution.

A related aspect of the first limitation concerns the number of cases investigated in this research—five instances of technical outcomes that greatly exceeded the initial expectations of the management and one that just barely met them. Because of the small number of projects that are investigated, these results are suggestive rather than definitive. Furthermore, in the absence of studies of "failed" projects, the combined structural effect and the design mechanism cannot be conclusively affirmed; technical projects that represent both clear-cut successful and failing outcomes will need to be studied. The literature on project management and teams as well as Ragin's qualitative comparative methodology would predict that more than one set of combinations of factors

is likely to produce the same outcome. Given the restricted number of projects that I studied, though, the variability of conditions was too low to afford their discovery.

Third, past research on technological innovation has demonstrated that each type, whether radical or incremental, is likely to be facilitated by different organization design and leadership style (Green, Gavin, and Aiman-Smith 1995; March 1991). Empirical studies of R&D projects, however, have seldom approached the investigation in this manner. In summarizing the literature on project management, Shenhar (2001) concludes that projects are by and large treated as if "one size fits all." Recent scholarship has called for empirical investigations to differentiate between R&D projects that represent radical innovation and those that represent incremental innovation (Astebro 2004; Balachandra and Friar 1997; Shenhar 2001). To this end, understanding the causal mechanisms that underpin the success and failure of technologically innovative projects will greatly improve with the investigation of the combination of factors and design choices that could explain the outcomes in cases that represent radical as well as incremental technological innovation.

For instance, judging from the technical and financial impact that each project had (as described in Appendix 1), it is conceivable to conclude that the five "highly successful" cases represent technologies and processes that could be placed at various points on a continuum of "exploration of new technical possibilities" as opposed to "the exploitation of old certainties" (March 1991: 71). If that should be correct, it could be argued that my findings speak louder to projects of a radical technical nature, while the proper management of projects that are more routine could be secured through a different combination of critical success factors and structuring of positions and role assignments (Dobrev, Kim, and Solari 2004). Much to my own regret, though, as I have acknowledged in Chapter 2 and Appendix 2, in the course of the research I had to abandon my initial intent to classify the six projects along these two dimensions, as it proved to be both practically difficult and wrought with ambiguity.

A fourth limitation stems from the fact that these findings are derived from qualitative and social networks data gathered at one point in time. As a consequence, idiosyncratic aspects of particular projects and their dynamics cannot be ruled out. Last, all six projects are drawn from the same organization. While

acknowledging that the nature of the innovation process in general does not change drastically from company to company, and that the size of each project at Global East's R&D laboratory is average for the industry, it is plausible that the findings are specific to this particular organization. For instance, at Global East it is the laboratory director who initially puts project teams together. R&D organizations that have different practices of selecting teams might give prominence to other types of social networks and positions and, ultimately, construct different narratives of success. Therefore, a large-scale, hypothesis-testing research study, involving multiple R&D organizations and many technical projects, will be necessary before drawing definitive conclusions.

Notwithstanding these limitations, I would argue that the very nature of the innovation process in general does not change from organization to organization. On the basis of empirical evidence from the innovation literature and on insights from the pilot study, I would contend that the way in which task sequences are put together in Global East is not significantly different from other R&D organizations. As just noted, the size of the projects, as well as the size of the laboratory, are average for the industry; the decision-making processes and constraints as well as the technological stages that each project at Global East goes through are similar to the ones described in most of the innovation literature.

In conclusion, I am confident that my findings are pertinent to the study of project management of technological innovation and will be informative for technology managers. While not all technical organizations are aligned to the traditional concept of research and development, without a doubt they too will recognize many of the daily challenges faced at the R&D laboratory at Global East as their own. One such major and emerging issue is that of project and organization design forms that will create a fluid, and yet dependable and predictable, system of capturing, retrieving, and sharing critical knowledge and information across a spectrum of stakeholders. The importance of generating insights into this matter will only become more crucial at a time of higher uncertainty, ever-increasing volumes of knowledge and information, paralleled by narrow specialization and the prevalence of flat organizational forms with fewer opportunities for traditional promotions. I discuss those in more detail in the last part of this chapter.

THE IMPLICATIONS OF THE LESSONS

With the just-discussed limitations in mind, I revisit the main findings of my research and discuss the theoretical and practical implications for the study of technological innovation and R&D project management. I begin by discussing what my findings mean to theories of organizations, project management, and small-group studies.

Theoretical Implications

Organizational Theories and the Literature on Innovation

Despite a prior recognition in the literature and some empirical research on the ways in which the formal and the informal structures influence each other, the interplay between the two, and the manner in which it affects performance and outcomes, is not well understood (Cross and Sproull 2004; Smith-Doerr and Powell 2005). To organization theory, approaching the understanding of performance and outcomes as being shaped through the interpenetration of the formal and the social network structures at multiple levels, and in particular through positions of centrality and roles, offers a fresh way of looking at classical structuralist explanations (Giddens 1979). Specifically, the examination of the structure-agency interaction through the investigation of the R&D organization and the projects' formal and informal structures and positions of power and prestige, as well as the cultural norms, organizational processes, and assignment practices, opens a window into the understanding of the mechanisms that shape the dynamics as a result of which structure shapes and is being shaped by the human agent in a way that enables the constitution and the reconstitution of the meaning, morality, and power conducive to the production of new technological knowledge. Scholarship has called for taking the theory of organizations to a new level by ending the nearly century-long "tendency in the literature to oscillate between viewing the formal and the informal organization as most significant. . . ." (Nohria and Gulati 1994: 529). "What we need," the authors continued, "are theories that focus on how the formal and the informal structures of an organization are interrelated and influence each other." By studying both structures simultaneously and examining the specific ways and processes through which they interpenetrate and

reconstitute one another, and shape outcomes, this research has responded to their appeal in an empirical investigation from a design perspective at the R&D project level.

In so doing, the analysis speaks to what Starbuck and Nystrom declared to be imperative about organizational design: that "it should mix simultaneous solution attempts. . . . A well-designed organization is not a stable solution to achieve, but a developmental process to keep alive" (1981: xx). Such a "double helix" structuralist approach provides opportunities not only to describe action in a social context but also to understand how it came about. This is achieved by looking at the variety of processes and mechanisms that are shaping and being reshaped by the interaction between the components of both structures. The need for approaching the study of technological innovation in this manner at the levels of the group and the project is even more warranted. The dynamics that emanate from either structure on its own, or the two together, through positions, statuses, and roles and their sequencing, are arguably more intense and complex at this level given the size of the group, the proximity of the actors, and the frequency of the interaction between them imposed by the nature of the research and development. Furthermore, both structures have been shown to play a crucial role in the transfer and sharing of knowledge and information, in the absence of which technological innovation is unthinkable. By addressing the question of success on technical projects through the investigation of the simultaneous and combined effect of both structures from a design perspective, the book makes several specific contributions to the literature on technological innovation and social networks.

My findings unveil a unique way of allocating tasks, responsibilities, and authority on technical projects. It contains the potential to reinforce the beneficial sides that each structure has to offer to performance while suppressing, to a degree, their limitations. In Table 5.1, I summarize the findings from prior research on the positive as well as negative effects of structure, both formal and informal, as well as the impact of human and social capital on outcomes.

As shown in the table, the formal and the social network structures, as well as the human and social capital, carry the potential to exert both positive and negative effects on the process of technological innovation and its outcome. Moreover, the formal and informal structures are often said to yield opposite

Table 5.1

Prior research on the effects of structure and the impact of human and social capital on outcomes

Element	Benefits	Deficiencies
Formal Structure	• Predictability • Accountability • Decision-making chain • Clear assignment of roles and responsibilities (Blau 1955; Blau and Schoenherr 1971; Burns and Stalker 1961; Galbraith 1973; Hage and Aiken 1970; Perrow 1967; Pugh, Hickson, Hinings, and Turner 1968; Scott 1992).	• Slow and redundant information channels • Slow action and problem solving in fast-paced environment • Stifled creativity • Duplication of work • Reduced learning opportunities as a result of narrow task specialization (Blau and Schoenherr 1971; Burns and Stalker 1961; Hage 1980; Hage and Aiken 1970; Perrow 1967; Scott 1992).
Human Capital	• Creativity (Amabile 1988) • Problem solvers (Becker 1964; Lin 2001)	• Could be disruptive and destructive; promotes the "Prima donna" complex • Could be hard to incorporate into the formal structure • Could create tension between the formal and informal authority and thus reduce the legitimacy of the formal leaders.
Social Networks and Social Capital	• Efficiency by coordination of critical task interdependencies (Coleman 1990; Gargiulo and Benassi 1999) • Reduced duplication of information and uncertainty that results from task interdependence (Blau 1955; Gargiulo 1993; Pfeffer and Salancik 1978) • Access to information and speeded up information exchange (Gargiulo and Benassi 1999; Podolny and Baron 1997; Lin 2001) • Learning opportunities (Lin 2001; Podolny and Page 1998; Powell and Brantley 1992) • Economic benefits thorough lowering the transaction costs (Uzzi 1997; Williamson 1991) • Increase of the respect, trust, and trustworthiness between actors (Bourdieu 1986; Fukuyama 1995; Putnam 1993) • Attainment of status through association (Baum and Oliver 1992; Podolny and Page 1998; Stark 1996;) • Maintenance of social norms, obligations, and expectations by social sanctions (Burt 1992; Coleman 1990; Granovetter 1985; Lin 2001; Putnam 1995)	• Need to be maintained constantly, as their instrumental value resides in the match between the needed and the available resources (Gargiulo and Benassi 1999) • Their maintenance could be burdened by "the particularistic demands posed by the same relationships purportedly responsible for the initial success of this same actor" (Gargiulo and Benassi 1999) • The same networks and social capital that are useful and enhance the positions and the performance of one set of social actors also carry the potential to hinder and even harm the ability of another social actor to pursue or achieve his or her interest (Coleman 1990; Portes and Sensenbrenner 1993) • Could lead to network closure, "relational inertia," and, consequently, unawareness of other possibilities and resources (Gargiulo and Benassi 1999; Grabher 1993) • They lack legitimate authority to "arbitrate and resolve disputes" (Podolny and Page 1998) • Dense social networks could have an oppressive effect especially in closed networks and in small communities (Brass and LaBianca 1999; Fischer 1980)

effects on technological innovation. For instance, while the formal structural arrangements secure the clear assignment of roles and responsibilities, along with the legitimization of authority, the social networks and social capital are deficient in fulfilling either of these functions. Unlike the informal structure, the formal specifies clearly where the accountability lies, which provides for predictability and stability. This is, of course, rather important given that the vast majority of technological innovation takes place in complex social organizations, and very often in large business organizations that rely on bureaucratic means as a way of achieving the necessary coordination and efficiency. At the same time it is an implicit expectation that the slowness and redundancy of the formal channels will be compensated for by the flexibility and agility of the informal ones, given, of course, that the potential they carry for negative and destructive action is neutralized. A pertinent question then is, *How* can their elements be put together in a manner that circumvents their less desirable impacts, while reinforcing their beneficial sides and not leaving the positive influences to chance? In other words, how can one unravel the perennial organizational search for a balance between "too much and too little structure" (Brown and Eisenhardt 1998)? My findings offer some insights into this matter and into the structuralist approaches of organization theory. They suggest that there are advantages to be gained when the formal and informal structures are entwined by embedding human and social capital into formal positions and roles on the projects. I discuss these next.

The "Bright Sides" of Social Networks and Social Capital: How Are They Reinforced at Global East?

In addition to the well-established benefits, demonstrated by prior research, that each structure provides to a project's success (shown in Table 5.1), their interaction effect, as the study of the laboratory at Global East suggests, gives rise to further ones. This is because when human capital and advice centrality are channeled through positions and roles in the formal structure, the beneficial effects of the social networks and social capital are multiplied, while the potential for the deficiencies that each one of them carries is avoided. Among those benefits, five can be identified as particularly prominent and of critical importance to technological success. The interview transcripts offer abundant

empirical evidence to support each of these five ways to enhance success on innovative projects.

First, team members are not forced into organizational boxes that prescribe behaviors and responsibilities inconsistent with their human and social capital. In other words, the Peter Principle (Peter and Hull 1969) is circumvented, as people are not promoted to their level of *in*competence. This helps to minimize not only an individual's frustration, but also that of the people he or she must work with or manage, thereby creating a better overall work atmosphere.

Second, interweaving human and social capital into a project's formal structure legitimizes the authority on each project and within the company at large (Podolny and Page 1998). As Burt aptly points out, "legitimacy does not come with the job; it has to be established" (2000: 390). The incorporation of the human and social capital within the formal projects' structures legitimizes the authority of each project manager by eliminating possible discrepancies and tensions between the formal and the informal decision-making lines of command.

If, for instance, a laboratory member with low technical human capital occupies a formal position of high technical responsibility, and another member with low managerial human capital occupies a position of high responsibility, the basis of the authority of these individuals is likely to be at stake, for the "social capital is the contextual complement to human capital" (Burt 2000: 347). Coleman (1988) provides evidence in support of this point in his study of high school dropouts. There he found that if the human capital that is possessed by the individuals is not complemented by social capital embodied in social relations, it becomes irrelevant for the people who have acquired it, as they cannot draw benefits from it. By the same token, the formal structure is the rational-legal complement to the human capital, already supplemented by social capital. In other words, what is accomplished through the formal structure is the legitimization of the already established informal authority that the individuals with high human and social capital have earned.

Third, when centrality is not simply recognized in the technical and organizational advice networks but also prescribed through formal positions and roles, resources in the organization are deployed more effectively and the utilization of the laboratory and projects' resources is not only secured

but maximized (Lin 2001). This is accomplished by the ability to mobilize the two forms of capital—human and social—so that neither remains idle at any time (Allen, Tushman, and Lee 1979). This in turn provides better access to, and mobilization of, the financial capital as the histories of the projects demonstrate (Granovetter 1985; Portes and Sensenbrenner 1993; Uzzi 1997, 1999). As Mote notes, "it is important to determine not only what the project members *know*, but also what they *do* within the context of the project" (2005: 109). Knoke has shown that "just as an individual can mobilize her personal contacts' social resources for purposive action, so can a formal organization activate various resource networks to achieve its goals" (1999: 20). What is more, he defines "corporate social capital" as "processes of forming and mobilizing social actors' network connections within and between organizations to gain access to other actors' resources" (1999: 17). At Global East's research and development unit, people with a lot of human capital who also occupy positions of centrality in the work-related advice networks—Fred, Nick, Olga, Tom, Natalie, and Ted—are highly visible within the laboratory as well as on the projects in which they are members. They also occupy key positions in either the formal laboratory structure, the formal structure on projects, or both, and their legitimacy transcends those boundaries because their expertise and services are recognized and often requested by other organizational units. As such, the management at Global East ensures the full mobilization of the available human and social capital while avoiding the duplication of roles and responsibilities. The result is that the overall social capital of the laboratory grows, which then increases the likelihood of corporate support on future projects.

Fourth, converting centrality in the two work-specific advice networks into formal centrality on projects provides for the rapid and high-quality exchange of information between the *relevant parties*, as opposed to just between *any parties* (Burt 2000; Uzzi 1996), and, hence, it improves the decision-making process at the individual, project, and laboratory levels. As Coleman (1988) suggests, information channels are the basis for any social action. Ultimately, technological innovation and its management are understood as knowledge and information-processing activities (Clark and Fujimoto 1991; Kerssens-Van Drongelen, Weerd-Nederhof, and Fisscher 1996; Tushman and Nadler 1986;

Utterback 1974). It is only logical to conclude then that "project effectiveness would be a function of matching communication patterns to the information processing demands of the project's work" (Tushman 1978: 640). Put differently, the preeminent aim of the structural design, in any organization dedicated to innovation, is to ensure the gathering of quality information and its most efficient and targeted redistribution, utilization, and exchange.

The issue for the R&D managers, therefore, is to create a business model that identifies, captures, and processes the relevant knowledge and information in a way that results in successful outcomes. Embedding human and social capital into the formal structure attends to this process requirement and, thus, provides for increased speed and enhanced quality of the information exchange between the relevant parties (that is, the actual linkages). It maximizes the utilization of the organizational and project resources because the formal and the informal decision-making channels overlap, since the people who are "stars" in the social networks, and whose advice is sought informally, are assigned to formal positions of authority that correspond to their technical, organizational, and social expertise. In this way, critical resources, including time, money, and man-hours, are conserved, and performance improves. Lin's (2001) theory of social capital, and the results from studies by Nahapiet and Ghoshal (1998) and Tushman and Scanlan (1981), back this conclusion. For instance, Tushman and Scanlan found that

> perceived competence is a more powerful determinant of internal consultation than is formal status. Technical competence and status are related. . . . [Technical] competence is associated with colleague consultation and may, in turn, be associated with a promotion. Formal status then further facilitates an individual's ability to serve as a boundary spanning individual. Individuals consult those whom they see as competent, independent of formal status (1981: 302).

That the entwining of the formal and informal structures through positions of centrality and roles could be beneficial for the exchange of information in the innovation process is also supported by Burt's finding that a "leader with strong relations to all members of the team improves communication and coordination despite coalitions or factions separated by holes within the team" (2000: 393). Most recently, Balkundi and Harrison have found that the "leader centrality is positively associated with team task performance" (2006: 59).

Last, along with legitimizing authority; maximizing the use of the human, social and financial capital; and speeding up the exchange of information between the relevant parties while increasing its accuracy, the incorporation of human and social capital into the formal structures of the projects as a design mechanism sustains, and in turn is sustained by, constructive organizational culture. The norms, values, practices, and organizational routines of such culture are also conducive to a high level of individual and group learning and performance in a knowledge-intensive work environment. By having members work on multiple projects and play different roles, the design promotes perpetual learning, both technical and social; allows laboratory members to develop worldviews that are not confined to one position only; and promotes the sharing of knowledge. These were discussed in more detail in Chapter 3. Hence, the project design matches the laboratory's cultural norms and values. In so doing, it is conducive to technological success, as it reflects the nature of the work in the R&D lab and the structural design requirements that stem from it.

The "Dark Sides" of Social Networks and Social Capital: How Are They Circumvented at Global East?

The positive aspects of social networks and social capital have received an enormous amount of attention in the sociological and managerial literature. During the past three decades these have been carefully analyzed, articulated, and documented. As is the case with most social constructs, though, their negative features have caught less attention and received less coverage. It has been only recently noted that a shift is beginning to take place from examining absolutes to examining trade-offs (Brass, Galaskiewicz, Greve, and Tsai 2004). Nonetheless, there are a handful of researchers who have warned against the euphoria surrounding the themes of social networks and social capital, and pointed out some of the disadvantages associated with them (Labianca, Brass, and Gray 1998). Gargiulo and Benassi, for instance, express the view that "since the instrumental value of social capital lies on the match between the resources needed by an actor and the resources provided by the actor's contacts, changes in the actor's task environment may require changes in the composition of his social network" and thus their positive effects will be diminished (1999: 302).

Let us consider the situation of a researcher on a project who urgently needs help with running a test on a highly specialized testing machine. He is unclear how to go about it but he solves his problem by tapping the knowledge of an engineer who specializes in this type of testing. The engineer works for a different developmental group, but the two are social friends, and the engineer manages to squeeze in our researcher's testing needs into his cramped laboratory's schedule and runs the test for him. Let us now suppose that the very same researcher works on a different technical project. A different type of test must be run, but this time no one in the organization has the expertise to do it. According to Gargiulo and Benassi (1999), as our researcher's task environment has changed and his contacts can no longer match his needs, he cannot capitalize on his social network. Therefore, he will need to recompose his social network—an activity that is likely to take time while its outcome cannot be guaranteed. Thus, in the short run he would not be able to resolve his testing issue by resorting to his informal contacts.

At Global East this potential social network and social capital deficiency is addressed by assigning the organizational members to work on several projects and to simultaneously occupy multiple roles on these projects. This includes having individuals from other business units and functional areas on projects as well. As a result, each of them constantly changes his or her task environment, increases his or her human capital, and concurrently expands and maintains his or her social network. In addition, the management at Global East has put in place mechanisms for the reproduction of the work environment on the projects by constantly and deliberately reproducing the organization's human and social capital. In this manner, the organization's survival and success in the long run are secured. How is this being accomplished?

The reproduction of social and human capital within the laboratory secures access to corporate financial capital and eases its mobilization. What is more, the routine of training and mentoring (Higgins and Nohria 1999) is not confined to the boundaries of the laboratory. The facility trains specialists from other business divisions as well as freshly minted university graduates who stay with the company but not necessarily with the laboratory. Thus, both the task environment and the social networks within the laboratory and outside are continually reconstituted. The constant changes in the role and responsibility

assignments, and the cultural norm of assistance, guard against the "prima donna" complex and the development of small, dense social networks within the organization that could lead to closure, duplication of the information, and the loss of awareness of opportunities and resources. It also helps to avoid what Levitt and March (1988) so fittingly termed the "competency trap."

Thus, by the blending of the formal and informal structures on the R&D projects the decision-makers at Global East have tackled two crucial issues in the management of technological innovation: the reduction of uncertainty that stems from task interdependencies and the increase of adaptability and flexibility at the individual, project, and organizational levels (Gargiulo and Benassi 1999). Hence, such project design enables the use of fluid yet specific structural mechanisms, including social networks, for aligning experts with organizational goals and challenges.

The empirical findings from this study challenge the prevailing views with regard to the relation between formal and informal structures on project design and outcomes. They demonstrate that on the successful R&D projects the two structures shape the outcome not in parallel but in interaction. Therefore, to examine the effects of the two structures separately, or to even conceive of them as either opposing or merely absorbing each other's inefficiencies, would be not only an artificial exercise, but also a barren one (Blau and Scott 1962).

Social Networks and Innovation

A further contribution to the study of innovation and project management comes from the application of the social networks perspective to technical environments. Specifically, it is reflected in my investigation of the effects of two complementary work-related advice social networks, pertinent to a technological environment. On the basis of the information-processing approach, prior research has firmly established that both internal and external technical communication are vital to the success of R&D projects. While building on this stream of empirical studies, my research extends it by offering a conceptual as well as a methodological development. The conceptual innovation is reflected in the distinction between two different types of information needs and exchanges that are specific to a technical environment—seeking advice and information on technical matters, and seeking advice and information on

organization and coordination issues. The creation of the two advice networks which capture the content of knowledge and information exchanges that are specific to R&D work processes—technical and organizational—is at the core of this formulation. My study is the first one to draw a distinction between these task-related complementary advice networks and to investigate the effect of centrality in both of them for the outcome of technical projects (Rizova 2002, 2006a).

The main focus of the social networks studies and the literature for the past three decades has been on the structure of networks, to the neglect of the importance of the type of ties and content (Adler and Kwon 2002; Nebus 2006). The majority of the social networks research on innovation takes a broad approach and looks at their effects in general (Adler and Kwon 2002; Cross and Sproull 2004). This does not capture the intensity and multifaceted nature of interpersonal networks, and as result fails to investigate the complexity of their impact on the success and failure of technical projects (Oliver and Liebeskind 1998).

From studies on transactive memories we know that people in organizations develop knowledge of "who knows who and who knows what" (Wegner 1995). It is only logical to expect then, that individuals will seek advice on different matters from different people within the organization. For instance, Cross and Sproull found that in knowledge-intensive environments professionals were in need of receiving a combination of five components of what the authors call actionable knowledge: solutions, referrals, problem reformulation, validation, and legitimizations (2004: 449). No one has received all five components from the same source; the highest number that one relationship provided was three out of the five. Hence, the separation of social networks in terms of technical and organizational advice has proven extremely helpful to illuminating the dynamics of technical projects, and both measures have been shown to contribute to the explanation of the projects' outcomes.[2]

A further contribution of the book to the literature on social networks is reflected in showing a process through which social networks and, more important, work-related advice networks are being generated. By doing so, my research speaks to the most recent call in the literature on social networks to start paying close attention to two important issues: the generation of advice

networks and the motivation behind it (Nebus 2006). In addition, by look-ing at the combined effect of both structures at a group level, the study offers useful insights into understanding how groups, not just individuals, acquire social capital.

Insights into these questions are critical to the study of innovation, as they shift the attention away from a bias in the social networks literature "toward examining a network's structural characteristics to predict outcomes" while not downplaying their importance (Nebus 2006: 615). In consequence, my study addresses what Krackhardt and Hanson (1993) called a point of cen-tral importance in the analysis of the effect of social networks on outcomes— namely, identifying the *critical types of networks* in an organization. As the findings show, this distinction helped to reveal the dynamics on technical projects in terms of the flow of knowledge and information, the coordination of critical task interdependencies, and the reduction of uncertainty. Further-more, by examining and identifying the ways in which centrality in the two advice networks is incorporated into and shapes the innovation process, my study contributes to the literature on social networks by speaking to Powell and Grodal's 2005 call for future research to focus on the specific ways in which social networks affect innovation.

Finally, although the empirical interest in the effect of social networks at the interorganizational and intraorganizational levels has been plentiful, as I have shown in Chapter 1, less than a handful of studies of innovation have employed social networks at the R&D project level; my study is one of those.

Organizational Learning and Innovation

By looking at repositories of knowledge, organizational routines, norms and practices, and the ways in which formal and social network structures affect the acquisition, sharing, and utilization of technical and organizational knowl-edge, the book speaks to the literature on organizational learning. Specifical-ly, it provides insights into how individual knowledge and learning could be transferred to and utilized at the level of the project and the organization. It identifies and describes the mechanism, processes, procedures, norms, and routines through which the organization imparts knowledge on its individual members. In so doing, the analysis tackles two of the classical debates in the

organizational learning literature that stand out as either unresolved or only partly answered. The first considers whether organizational learning implies behavioral or cognitive change and how to reconcile the two. The second seeks an explanation for the linkage between the individual and organizational levels of learning. What is more, this study is one among a handful that discusses learning at the level of the group.

The first debate in the field of organizational learning looks at the content of organizational learning and adaptation. In this regard, a distinction has been drawn between cognition and behavior. Fiol and Lyles (1985) depict the difference in a sense that *learning* reflects changes in cognition, whereas *adaptation* reflects changes in behavior. Cognition affects action and vice versa. In other words, it is a difference between gaining new knowledge (cognition) and putting it into use (behavior). The tension between these two aspects of learning comes as a result of the fact that cognition and behavior do not necessarily occur in parallel. It is plausible that changes in behavior may take place without the development of cognitive associations and changes. It is just as true that learning may or may not lead to changes in behavior or organizational performance; what individuals learn and know may not even be extracted and shared for the purposes of the organization. For instance, small and incremental behavioral changes do not necessarily result in important learning.

At the same time, there is no empirical evidence that suggests that large-scale behavioral changes would lead to proportionally large changes in cognitive associations. Fiol and Lyles (1985) illustrate this point by using the example of the wave of mergers in the 1960s, when rapid and profound changes were taking place in the forms of acquisition and yet there was an absence of learning. When studying organizational behavior under the conditions of immense uncertainty and crisis, Starbuck, Greve, and Hedberg (1978) found that the firms' response was to keep introducing various changes in the hope that one would eventually work. The issue that scholars in the field grapple with is how, and in what ways, this tension might be resolved and the two perspectives be integrated. So far as technological innovation is concerned, this is an important element of creating a learning organization that has the structures and culture in place capable of acquiring knowledge, sharing it, and using it.

Knowledge relevant to technological innovation is not only complex but

widely dispersed (Powell 1998; Shan, Walker, and Kogut 1994). Yet identifying this critical resource and bringing it into the organization is imperative to technical success. Obviously, it is not sufficient to locate it and to bring it in, unless it is used. By specifying the mechanism of prescribing the actual advice relations that take place in the process of the development of new technologies, my analysis offers insight into how the gap between locating knowledge and using it could be closed.

My research would suggest that the project-design mechanism at Global East, coupled with the type of organizational culture and processes the company has developed to reinforce the design form, offers a way of bridging the two perspectives. Individuals who acquire technical and organizational knowledge and skills do not keep it to themselves; rather, the knowledge they have in their possession is extracted from them, and shared and transferred to the project, and ultimately to the organization. This comes through the effect of the project's design form, the cultural norms and expectations, and the organizational processes (Argote, McEvily, and Reagans 2003). Critical among those are the norm that compels lab members to help one another solve technical issues, the fact that performance evaluations and pay raises are based on yearly performance, and that individuals occupy multiple positions and play different roles on projects at the same time. Thus, members are constantly in interaction with each other and, more important, the interaction is purposeful and project-oriented. Assignments on future projects are based on an individual's reputation regarding the utility of his or her technical as well as social competencies and skills. Hence, Global East's employees have the incentive to share and use knowledge, rather than to withhold it, to learn broadly while developing a narrow specialization at the same time.

Second, learning in organizations presupposes that individuals gain knowledge and that what they have learned is retained not by the individual only but by the organization as well. This, the literature suggests, is accomplished by creating routines. These are not developed by the organization, however, but by its individual organizational members, who create and carry them out. Routines acquire a life of their own, though, and they endure even when their creators leave the organization. Individual learning, therefore, is a necessary, but not a sufficient, condition for organizational learning. Hence,

organizational learning theories would suggest that institutional processes must be put in place to record, store, and transfer what has been learned by individual members to the organization, and then back to all organizational members. What is more, there is a consensus in the literature that organizational learning is not just the cumulative knowledge possessed by individuals (Fiol and Lyles 1985). Thus, any model of organizational learning must be a cross-level model and needs to provide an answer to the questions about how individual knowledge is shared and how the organizational knowledge, codified in routines and the firm's culture, is transmitted to new and existing individual members (Crossan, Lane, and White 1999). Although the significance of formal and informal organizational socialization processes, through training and mentoring, organizational rituals and ceremonies, and myths and storytelling, has been well documented and acknowledged, a definitive answer to this question has yet to be found.

By closely examining not only the project-design mechanism but the various elements of the organizational culture that support it, together with the managerial practices in place, my findings shed light on the second debate in the literature on organizational learning. More specifically, this research contributes to an explanation of how knowledge is transferred from the individual to the project, then to the organizational level and back, and ultimately is utilized in the creation of new products and technical processes. I suggest that this builds a perpetual learning environment; one in which the creation of new knowledge and the storage of the old wisdom is rewarded. In examining this, the book also speaks to the classical question in the social sciences concerning the interaction between structures and human agents (Giddens 1979). The role-assignment practices at the R&D laboratory offer an insight into a mechanism that could match structures and agents in a way that enables the continual constitution and reactualization of the meaning, morality, and production of new technological knowledge.

Intraorganizational Learning, Project Management, and Small-Group Studies

The results also help to shed light on some of the unresolved questions in the literature on small groups and organizational learning (Argote and Ophir

2002). McGrath and Argote (2001) developed an analytical framework in which they creatively link two explanatory approaches that have a significant bearing on my research—that of organizational learning and that of small-group and projects studies. In particular, their framework connects three subprocesses of organizational learning—creating, retaining, and the transfer of knowledge—to the three basic elements of any small group and, for that matter, organization—"members-tools-tasks" (Arrow, McGrath, and Berdahl 2000). The combinations of these three elements can be described as seven subnetworks: (1) the member-member network (describing the social network of the organization); (2) the task-task network (which represents the sequence of purposive actions, tasks, and routines that are employed in an organization); (3) the tool-tool network (depicting the types of tools and technologies used by the organization); (4) the member-task network (describes who does what); (5) the member-tool network (showing who uses what kinds of technologies and tools); (6) the tool-task network (demonstrating which tools are employed in the process of accomplishing which tasks); and (7) the member-task-tool network (which points to who does what and with which tools in an organization).

Based on this framework, McGrath and Argote (2001) and Argote and Ophir (2002) argue that an organization's performance will improve with the increase of both the internal compatibility of the networks and their external compatibility with other networks. "For example, when the member-task network allocates tasks to the members most qualified to perform them, the member-task network is internally compatible and organizational performance improves. Similarly, when members have appropriate tools to perform the tasks allocated to them, performance improves because the member-task network is compatible with the member-tool network. Thus, according to this framework, performance increases as the compatibility of the various networks increases" (Argote and Ophir 2002: 182–3). Empirical research has generated support for the argument regarding the relationship between the compatibility in various subnetworks and performance (Hansen 1999; Hargadon and Sutton 1997). For instance, Wegner (1995) has shown that groups with well-developed transactive memory systems (that is, the combination of group members' individual knowledge and their awareness of who is in the

group and who knows what) perform better than those lacking this quality. More empirical research is needed, though, to provide support for this framework for each subnetwork. Argote and Ophir (2002: 200) called for further exploration of "how the member-member network, the member-task network, the member-tool network and the member-task-tool network affect the creation, retention, and transfer of knowledge within organizations." Through the investigation of members' human and social capital, as well as the specific ways in which the organization makes use of both, my study provides an empirical case that supports the fundamental tenets of this framework and the member-task-tool network in particular.

Furthermore, I approach the understanding of outcomes on projects from both a holistic and a dynamic perspective. Not only do I look at how structures, positions, and roles affect success, but I also take into account how the unique allocation of tasks, responsibilities, and authority is reinforced by the organization's culture and supported by specific processes. In contrast, the vast majority of studies at the project level had investigated the effects of structures, climate, and processes on innovation separately (Anderson and King 1993). What is more, I adopt McGrath, Arrow, and Berdahl's (2000) view of projects as complex, adaptive, and dynamic systems that are subjected to the simultaneous impact of local, group-based, and contextual dynamics. In so doing, while my focus was on the projects *per se,* I have not investigated them outside of the organizational context within which they are embedded. By paying close attention to the histories of the projects, to the personnel assignments and the changes in those during various stages of the projects' development, and to the laboratory's culture and processes, I have moved away from attempting to understand groups in a static and linear fashion (Arrow, McGrath, and Berdahl 2000). Thus, my contribution is in offering one of the few empirical cases conducted from this newly emerging perspective for studying small groups.

Finally, the study speaks to a long-standing concern in the organizational behavior literature: namely, the need for a multilevel integrative investigation. I approach the exploration of technological success from the perspectives of the individual, the project, and the laboratory simultaneously. In the project management and team studies, the call for such investigations has been promi-

nently documented in the past several years, as the literature is changing focus from the long-standing static and linear "input-process-output" paradigm to dynamic and multilevel models. It has been argued that "multilevel models provide a deeper, richer portrait of organizational phenomena (Klein, Tosi, and Cannella 1999: 243) and allow even more integrated inquiry (Kostova 1999: 320)" (Gnyawali 2001: 435). Approaching the issue from these levels afforded a more comprehensive investigation of how the individuals' structural properties shape, and are being shaped by, the project and organization's properties and systems. It afforded a more coherent investigation of the projects' dynamics that are created and re-created in this process and how they influence the outcomes of projects. Finally, it has helped me to address the issue of how the meso and micro levels are connected and shape one another.

Implications for the Managers of Technology

The findings also carry significant practical implications. They uncover, describe, and analyze organizational and project management practices at an R&D laboratory of a *Fortune* 500 company that have clear relevance for other knowledge-intensive technical settings. The book identifies four critical success factors, a distinct project design strategy, and an organizational culture that put these factors together in a manner conducive to positive outcomes on innovative projects. In particular, the results will be informative to technology managers both for practical reasons and to enhance their understanding of the interactive relationship between structures and human and social capital in a technical environment. To that end, this work offers insights and lessons that go beyond simple benchmarking roadmaps.

An appreciation of the specific ways in which formal and social network structures interact on technical projects can help companies design and maintain learning organizations in which members exchange knowledge and information efficiently and willingly. Such sharing is becoming increasingly important, especially in knowledge-intensive industries in which the volume of information can increase exponentially over time. Together with this growth is the ever-increasing knowledge specialization and division of labor. Such trends impose substantial demands on companies to organize their laboratories in ways that maximize the acquisition, distribution, and utilization

of information that is critical to the survival of the organization. Accordingly, one of technology leaders' greatest challenges concerns the management of human and social capital and the manner in which they can be aligned with the organization's goals in specific yet flexible and fluid ways. A recent article on "thinking for living"" published in *The Economist* ("A Survey of the Company" 2006), which discusses the pressure to relentlessly search for innovative design forms that could address the needs of the emerging knowledge industry, states the following:

> "There are three broad approaches to knowledge management. One is to create a system where all information goes to everybody, which is hugely inefficient; the second tells people what others think they need to know, which may not match their real needs; and the third enables them to find for themselves whatever they want to know. Companies like to say that they aim for the third approach, but they do not always find it easy" (p. 10).

It seems as though the R&D lab at Global East has something to offer to this conundrum. Furthermore, it has been well-established that the "more that workers interact with each other, the more likely they are to solve the problems of complexity that are a feature of modern organizations," and to generate new ideas ("A Survey of the Company" 2006: 15). In the view of many experts in the field, the way to address the challenges that organizations face in the twenty-first century is to create organizations with structures and cultures that both promote and support interactions. The relevance of my findings to this impending need is obvious.

The stories of the six projects, and the lessons that can be drawn from them, speak loudly to yet another pressing issue in contemporary knowledge organizations. Flat structures are the predominant way of organizing the work in such companies at the present time. As pointed out in *The Economist,* flat organizations emerged in response to the external and internal challenges that the organizations of the twenty-first century face. Along with their beneficial sides, they have produced some unexpected difficulties for motivating the knowledge workers, as there are fewer and fewer opportunities for recognition through traditional promotion in these structures. This is further exacerbated by a growing degree of knowledge specialization. If organizations do not

find a convincing way of addressing this challenge, the managers' fear that the people with most knowledge and experience will look for traditional forms of recognition elsewhere, and take with them the knowledge, competencies, and skills in their possession, is a very real one. Through interweaving, structure, culture, practices, and incentive systems, the case of Global East offers rich insights to the managers of technology into this dilemma.

SUMMARY OF THE PROJECTS

Project	Number of Members	Project Manager(s)	Project Duration	Technical Objectives and Technical Outcome
Alpha	5	Fred	9 months	*Objective:* To modify an existing product and improve its performance. *Outcome:* New product; its commercial impact is not clear.
Beta	7	Tonya and Olga	1 year	*Objective:* The development of a new technical product, its successful testing and commercialization. *Outcome:* Threefold: development of a new product, a new measurement methodology, and a training component. In addition to its intended use, the product is widely used by the salespeople. It led to the establishment of a new manufacturing business.
Gamma	8	Fred and Nick	2 years, 3 months	*Objective:* The development of a radically different test equipment and process platform. *Outcome:* A new product and process platform. The product represented a significant change in the way an industry was operating, and Global East captured about 90 percent of the respective market.

Project	Number of Members	Project Manager(s)	Project Duration	Technical Objectives and Technical Outcome
Delta	7	Fred and Nick	3 years	*Objective:* To design a low-cost machine and process methodology that could be used in the company's facilities worldwide. *Outcome:* A new piece of equipment with unintended range of product-development capabilities for another business unit.
Epsilon	10	Natalie	3 years, 6 months	*Objective:* Designing and building an "R&D tool" for speedy and reliable data collection. The technical specifications involved reducing the size of the equipment, improving the speed of operation, and increasing the degree of user-friendliness. *Outcome:* Superior technical product.
Zeta	12	Ted and Tom	3 years	*Objective:* To develop a new technological process to increase sales to technical customers. *Outcome:* Technology platform that is used across the company's core business divisions. A second phase of the project has developed as a response to the customers' needs for technical support, equipment modifications, and training.

APPLYING QUALITATIVE AND QUANTITATIVE METHODOLOGIES TO THE STUDY OF TECHNICAL PROJECTS

"Most institutions demand unqualified faith; but the institution of
science makes skepticism a virtue."
Robert Merton (1938: 334)

INTRODUCTION

Officially, this research started with the support from a National Science Foundation (NSF) grant in the summer of 2000. My interest in technological innovation and the work of scientists and engineers in research and development laboratories, however, predates it. It began several years earlier with a summer internship at the Corporate R&D at the headquarters of the Gillette Company when I was a Ph.D. student. The internship turned into a rewarding three-year work experience. The more time goes by, the more I realize how extremely lucky I was to get to work for the vice president of Corporate R&D, Dr. John B. Bush Jr.—a true Renaissance man with a passion for learning. Our partnership grew and, along with three other colleagues from academia, we submitted a research grant to the National Science Foundation to investigate how companies organize for innovation. The rest is in this book.

When I started the research my aim was twofold. I sought to understand the ways in which both the formal and the social network structures, and their interaction, affected the outcomes of R&D projects. I also wanted to illuminate the combinations of the factors conducive to a successful outcome for technical projects and, provided that there was enough variety, to be able to investigate projects that exemplify radical and incremental innovations.

The Selection of the Study Site

Technological innovation, as I argued in Chapter 2, is understood both by academics and practitioners to be the result of a successful transfer of an idea, which comes from various types of knowledge and sources of information, into a new product or process that has a social and a market value (Kerssens-Van Drongelen, Weerd-Nederhof, and Fisscher 1996). Ideally, to achieve a thorough understanding of an instance of innovation, one needs to follow it along the lengthy path from the generation of an idea to its development and materialization in the shape of a new product or process, and, finally, to its commercialization. The research and development organizational function is the most technically essential and complex information-processing unit that is involved in the innovation process (Allen 1977; Utterback 1974). An attempt to investigate the phenomenon beyond an R&D laboratory in a single study would result in a less than comprehensive picture of the important dynamics that take place at the locus of technical creativity. These dynamics have an indelible effect on the other organizational units, without whose involvement innovations could not be completed. Furthermore, given the size of the study budget and the personnel, approaching the investigation from an all-inclusive perspective was simply impractical. While acknowledging that more in-depth understanding of an instance of technological innovation would benefit from expanding the research effort into the realms of marketing and manufacturing, the focus of this study is the research and development unit only. As part of an NSF project on the management of technological innovation, our research team secured entry into a Global *Fortune* 500 company located in the United States, "Global East."[1] We gained access to the R&D laboratory through an indirect contact that one of the team members had with the company.

The Selection of the Projects

The laboratory director selected the six projects discussed in the book on the basis of the following criteria. Each project (1) had to employ at least five project members; (2) had either been completed over the past two years or was taking place at the time of the study and was scheduled to be completed; (3) was initially estimated to last for one year; and (4) had to represent either a "successful" or a "failing" outcome. The six projects that met these selection criteria are the ones analyzed in this book.

The Selection of the Respondents

As the research design called for gathering data from all project members, I developed three comprehensive lists during the organizational mapping procedure: first, a list of all laboratory members; second, a list of all project managers on each selected project throughout its entire duration; and third, a list of all project members on each selected project throughout the project's duration.

The lists of project managers and members were initially composed by the laboratory director. In the course of the interviewing and questionnaire administration, I asked each project manager and member to verify the participation of the individuals on the list. During this procedure, one member was added to project *Alpha* and one was excluded; one member was removed from project *Gamma*; one was removed from project *Delta* and one was added to it; three were removed from project *Epsilon*; and five were removed from project *Zeta*. There were several reasons for excluding members from the projects' lists: (1) a member approached for an interview denied his or her participation on the project on the grounds that he or she was initially assigned to the project but shortly after was reassigned and therefore did not work on it. This was further confirmed by the rest of the project members; (2) three of the company's vice presidents were listed on projects on the grounds that they took a personal interest in the projects. They were not, however, involved in any capacity directly with the daily work, and therefore I eliminated them from the list; (3) organizational members who occasionally acted as informal consultants in their own time, availability, and desire were also excluded at their own request. They felt that they were not in a position to talk about

project dynamics for they saw their involvement as sporadic. Moreover, they did not participate in making any project-related technical or organizational decisions. By the same token, two individuals were included on two projects after their participation was confirmed with the projects' managers and members. In addition, one person who was listed on project *Zeta* strongly denied being a project member in any formal capacity and thus was excluded from it. There were no other discrepancies in the project assignment evaluations, and the final list was unanimously agreed upon. The number of members initially listed on each project, the changes made to the projects' lists, and the final list of members for each of the six projects under investigation are presented in Table A2.1.

The data for the study were collected from three sets of respondents. First was the director of the R&D facility. Second was every project manager on each of the six projects. Because some projects were assigned more than one manager, and some of those managers managed multiple projects, this set consisted of a total of ten interviews with seven individuals. Third was every accessible project member on each of the six projects, a total of thirty-two individuals. The population frame consisted of fifty individuals who were

Table A2.1

Project personnel numbers

Project	Number of Project Members Initially Listed	Number of Project Members Added to or Removed from the List	Final List of Project Members
Alpha	5	(+1) (- 1)	5
Beta	7	0	7
Gamma	9	(-1)	8
Delta	7	(-1) (+1)	7
Epsilon	13	(-3)	10
Zeta	17	(-5)	12
Total	58	(-9)	49

Table A2.2

Number and percentage of responses, by project

Project	Final List of Project Members	Number of Responses per Project	Percentage of Responses per Project	Reason for Nonresponse
Alpha	5	4	80%	1 not accessible
Beta	7	7	100%	-
Gamma	8	6	75%	1 refused 1 not accessible
Delta	7	6	85%	1 in India
Epsilon	10	8	80%	1 in Brazil 1 in South Africa
Zeta	12	11	91%	1 in Europe
Total	49	42	86%	7

listed and confirmed in their respective roles as a laboratory director, project managers, and project members on all six projects. Out of those, therefore, I gathered data from forty-three. The count and percentage distributions of responses by project are presented in Table A2.2. As the table lists the responses by project only, the answers of the laboratory director are not reflected in it.

THE COLLECTION OF THE DATA: SOURCES AND METHODS

Between June 2000 and January 2001, I collected qualitative and quantitative data on the six innovative projects in Global East's R&D laboratory. To ensure higher validity and reliability, the data were gathered and triangulated from six different sources. They are (1) *public sources:* the company's annual reports for the past five years, press coverage, and advertising materials; (2) *company's internal records:* the list of organizational members, list of projects within a specific timeframe, demographic characteristics, and members' organizational tenure; (3) *interviews:* a combination of unstructured and semistructured protocols; (4) *questionnaires:* a combination of open-ended and closed-ended questions; (5) *direct observation:* more than a dozen field visits to the site in

the course of nearly eleven months; and (6) *key informant:* the laboratory director of the facility, who championed the study and assumed the role of key informant. Three research team members met with the director on several occasions in the early part of 2000 during the site-recruiting efforts. The key informant was invaluable in providing us with an understanding of the business and political context within which the business division and the R&D laboratory operated. He was very accommodating and helped with making all the necessary arrangements for collecting the data. He also facilitated any additional contacts if and when needed. The high level of trust with which the director approached the investigation was, to a large degree, due to the fact that the vice president of Gillette's Corporate R&D was on the research team. Furthermore, Global East is in a different industry from Gillette, and hence, the two companies are not competitors.

In addition to using multiple data sources, I also sought a variety of types of information as a means of attaining richer information and providing for venues to check my understanding of the projects' histories and dynamics. Thus, I collected three types of data. First was *attribute data,* which tap into the attitudes, opinions, and behaviors of the social agents. Second was *relational data,* through which contacts, ties, or connections between individuals are identified. Last, I collected *ideational data,* which concern the subjects' meanings and definitions (Scott 2005).

The findings in the book are based on the analysis of the qualitative and quantitative data that I gathered from the three sets of respondents—the laboratory director, the project managers, and the project members. From the laboratory director I collected historical information and data regarding the company in general and the R&D facility in particular. Among the areas of interest were the number and types of divisions, lines of business, core business categories, profitability and growth pattern for the past five years, organizational leadership and leadership changes within the past five years, the number of employees, organizational culture, social networks, and the histories and outcome of the six projects. This information was collected through open-ended interviews that took place at the site on three separate occasions, each of which lasted for over two-and-a-half hours. The aim of the interviews was to achieve a general understanding of the company and the laboratory. Several

themes emerged from the first two interviews regarding the director's understanding as to why some projects exceeded the initial expectations, both technical and financial, and others did not. During the third interview, I focused the data collection on these emerging themes. In addition to the interviews, I designed a questionnaire for the position of lab director. Through this method, I gathered more specific financial and project information, such as reporting relationships, performance evaluation criteria, degree of procedural and financial discretion, and interdependencies with other organizational units.

The data from the projects' managers were also collected in two ways. First, I conducted open-ended interviews with every project manager. On average, the duration of each interview was an hour and a half. Three of the managers were interviewed twice to focus the data collection on the themes that emerged from the first interviews that pertained to their explanations about why and how success on the projects under their leadership was achieved. In addition to exploring the main research question with this set of respondents, the following major areas were also discussed in the interviews: the history of the project, its time frame, origin of the idea, the leadership and leadership changes, the project's structure and members, social networks, and the overall climate on the project and its outcome. Second, through the administration of questionnaires, which were prepared specifically for this set of respondents, I explored in more detail the communication patterns, style of leadership, decision-making processes, performance-evaluation criteria, rewards and recognition practices, and degree of technical and financial autonomy and discretion.

Last, the data from the projects' members were also collected in two ways: interviews and questionnaires. Initially, I conducted open-ended interviews with seven key project players, who emerged as such in my conversations with the director and the managers. Based on these open-ended interviews and on insights from the data collected from the lab director and the project managers, I developed a mostly structured interview protocol that explored issues specific to this set of respondents. An interview appointment was set up with each accessible project member, and the interview sessions lasted from one hour and fifteen minutes to an hour and forty-five minutes. The questions that I asked of project members were aimed at understanding

their experiences on the projects on which they participated. Among the areas I explored were the positions and roles they held on the project, whether these were assigned or voluntarily assumed responsibilities, their tenure with the project, the communication and decision-making patterns, the degree of procedural and technical autonomy they had, and questions about social networks and project outcomes. I also turned these interview protocols into a written questionnaire—a combination of closed and open-ended questions—and used it in two situations: when a project member was not easily accessible (for example, several people were on one-year assignments abroad) and when a project member who had worked on more than one of the six projects in my study had already been interviewed and there was a time conflict for scheduling a second or a third interview. In these circumstances, I asked the respondents to be as specific as possible and to write down as many examples as they could provide. A total of eleven responses were gathered in this manner. These respondents were handed or e-mailed the questionnaire form, along with a list of terms and definitions and a letter that explained the nature and the purpose of the study. Approximately two weeks before the questionnaire was sent to those who were abroad, the respondents were independently contacted by the laboratory director who, too, explained the purpose of the study and encouraged them to participate, while emphasizing the voluntary nature of their engagement. I then followed up with telephone interviews.

All interviews but one were tape-recorded with the consent of the interviewees.

MEASURES

After I reviewed the innovation and project management literature, and took into consideration the insights from two theories of structure—structural contingency theory and the social network perspective—I developed a set of measures that reflect the formal and the informal structures that have been shown to be relevant to successful innovation at the individual, project, and organization levels.

Project Outcome

The research design called for studying projects with "successful" and "failing" outcomes. In the R&D practice, however, there are no universally accepted measures of success and failure, particularly at the project level. In the absence of a matrix for classifying the outcomes of R&D projects, the scholars and industry practitioners have developed three answers to this question, summarized by Balachandra and Friar (1997). The most often used approach, in academia and in industry alike, is to ask a company manager or executive to classify a project's outcome based on a set of criteria the manager deems appropriate. This method is supported by an agreed-upon definition of performance as "the overall success of the new product as perceived by company management" (Meyer and Utterback 1995: 299). Sparrowe, Liden, Wayne, and Kraimer (2001) adopted a similar approach when they assessed the group performance of thirty-eight work groups in five organizations. Susanne Scott (1997), too, relied on this procedure in her study of forty-two product and process development teams. A second approach is for the researcher to make the determination after all data are collected and analyzed on the basis of a priori specified criteria (Keller 2001). These criteria are usually developed and grounded on insights from the literature. Last, a third option is to appoint an independent panel of experts who will make the decision based on developed industry-specific criteria (Einhorn 1972). It has been suggested, however, that this should be avoided (Astebro 2004). In this study, I relied on a combination of the first two approaches in conjunction with the use of the market criteria as a validation measure for the classification of the projects' outcomes as "low" or "high" success.

Following Keller (1994), I asked the director to classify the outcomes based on the degree to which each project met the management's expectations in terms of first, actual, and/or potential financial returns; second, satisfying the technical parameters of the project; and third, staying within or exceeding the budget constraints. These criteria, as the innovation management literature suggests, are the few composites, among many, often used internally in R&D organizations (Dvir, Raz, and Shenhar 2003; Keller 1994, 2001). The laboratory director felt that it was not justifiable to apply the dichotomous categories of "success" and "failure" to the six projects selected for the study. His strong

opinion was that a project's outcome is not a discrete category, but a continuum. As none of the projects were terminated prior to their completion, he argued that these projects represented "success" of varying degrees. Hence, I studied five "highly successful" projects and one "less successful."

Type of Technological Innovation

The types of technologically innovative projects of interest to me were *radical* and *incremental.* The distinction between them is based on the degree of change that is introduced to a final product or process when compared to the existing state of technology and the knowledge base (Damanpour 1988; Green, Gavin, and Aiman-Smith 1995; Utterback 1994). I asked the laboratory director and the project managers to classify each of the six projects as an instance of either radical or incremental technological innovation and provided them with the following definitions. First, radical technological innovation involves a drastic departure from the existing technology, knowledge base, and practices; it is normally linked to a breakthrough in knowledge (for instance, the introduction of the airplane, penicillin, or the computer). In contrast, incremental technological innovation corresponds to a little departure from the currently existing and relied-upon technical and scientific knowledge and practices; it only adds new features to an already existing product or process. It is based on the effective application of known methods and is systematically derivable from known technology.

The classification of the projects within these categories has proven, however, rather difficult, as the majority of my respondents felt that they could not draw a clear distinction between radical and incremental outcomes. In their opinion, the best way to describe the degree of technological change that each project represented was as a continuum. When I tried to apply a continuum to measure the category, the discrepancies and the nuances in the director's and the project managers' evaluations of the projects grew too big to regard the classification as valid. Such lack of consistency was further exacerbated by the limited number of projects in my study. As a result, I was likely to end up with six projects, the outcomes of which could be labeled as representing five degrees of innovativeness, and thus, with great reluctance, I dropped the category from the analysis all together.

The Independent Conditions

Altogether ten independent conditions are included in the study, derived from the review of the innovation literature and, in particular, the literature concerning structuralist approaches (Amenta and Poulsen 1994). Five of them reflect the projects' formal structure and five the projects' social network structure. The independent conditions that reflect the formal structure of the projects are degree of project centralization; degree of project task interdependence; degree of technical autonomy; the frequency of formal communication and reporting relationships between the project members and the project manager; and leadership style.

The informal social network structure in the laboratory was characterized by measures of network centrality. I adopted Ibarra's definition of social network centrality as "a measure that indexes centrality as a function of the centrality of those to whom an individual is connected through direct and indirect links" (1993: 480). Social networks data were collected from all lab members employed full-time, using both the realist and the nominalist approaches (Knoke and Kuklinski 1982). I asked each respondent four questions about his or her interaction with other co-workers to tap into four types of social relations that take place at the lab: instrumental, expressive, and two work-related advice networks that I developed to reflect the critical knowledge and information needs that are specific to R&D organizations (Rizova 2002, 2006a). The first one, *technical-advice* network, concerns seeking advice on technical matters. The second one, which I termed *organizational-advice* network, concerns seeking advice on organization matters (for example, scheduling, personnel and budget assignments, coordination of work activities, time management, and so on).[2]

On the basis of these four network questions, I developed the following measures, which I employed to characterize the social network structures: project's network density, project's network centralization, and individuals' structural centrality in each of the four social networks. The first two measures, density and centralization, are developed as network measures at the project level. The third one, structural centrality measured by degree (Freeman 1979), was developed and used at the level of the individual.

DATA ANALYTICAL STRATEGIES:
QUALITATIVE INDUCTIVE ANALYSIS, COMPARATIVE
METHOD, AND SOCIAL NETWORK ANALYSIS

My analytical approach with the qualitative data was inductive; I allowed the findings concerning the R&D projects' "high success" and "low success" to emerge from the narratives that the respondents created.[3] After all the interviews were conducted and transcribed, I began the data analysis by applying qualitative open coding to the transcript material that dealt with the respondents' answers to the questions asking them what, in their view, best explained the outcome ("high success" or "low success") of each of the projects on which they were members. I also analyzed their explanations and the elaborations and examples that they offered in their answers to the four social networks questions, as well as any additional comments relevant to their understanding of the outcomes and dynamics on each of the six projects. I first created broad open categories by putting together interview text from the transcripts relevant to these data sets. Then, following Strauss and Corbin (1990) and Lincoln and Cuba (1985), I began to refine these broad groupings and concepts into more focused categories and subcategories by examining in greater detail the examples, clarifications, and explanations my subjects provided. The final result of this process was the emergence of the four factors, which I discussed at length in Chapter 2, that appeared to explain the outcome of each of the six projects and the respondents' narratives on how each factor contributed to the outcomes. A distinct feature that came from this analysis was that the project managers and members saw these four factors as shaping the success on their projects together, in an interactive fashion.

To triangulate the findings from the inductive analysis of the qualitative interviews on the four critical success factors, and in particular their interactive effect, I employed Charles Ragin's Qualitative Comparative Analysis (QCA) (2000, 1994, 1987).[4] Given the small number of instances of technological innovation that I studied, as well as the limited diversity of projects in terms of the outcome, the QCA method is the most appropriate data analysis strategy to uncover the combinations of causal factors linked to a particular result and to test for interactive effects. In Ragin's words, the comparative approach is "especially well suited for addressing questions about outcomes resulting from mul-

tiple and conjectural causes—where different conditions combine in different and sometimes contradictory ways to produce the same or similar outcomes" (1987: x). This method is applicable to examining patterns of similarities and differences within a moderate number of cases. Methodologically, it provides a bridge between qualitative and quantitative research, as it integrates features of the "case-oriented and variable-oriented approaches" (Ragin 1987). It is based on, and makes use of, the ten principles of the Boolean algebra of sets and logic as a technique of qualitative comparison (for example, use of binary data, use of truth tables as data representation method, Boolean addition, multiplication and minimization, and combinatorial logic). Building on these principles, the comparative research method enables a researcher to address complex patterns of causation, that could appear even contradictory, while its methodology allows for the elimination of irrelevant causes (Ragin 1987). Like quantitative research, the causal factors and categories are specified at the onset of the research, and are based on a review of the literature and findings from previous research (Amenta and Poulsen 1994). Unlike quantitative research though, a comparative analyst would remain open to revisions of his or her framework. Also, the causal conditions (independent variables) are investigated in their conjuncture, and do not compete among themselves for the explanation of the variation in the dependent variable. Like qualitative researchers, "comparative researchers consider how the different parts of each case—those aspects that are relevant to the investigation—fit together; they try to make sense of each case" and systematically analyze the similarities and differences in an attempt to identify the patterns of diversity (Ragin 1994: 105). Thus, deep knowledge of the cases is of enormous importance, as is the reliance on qualitative data for checking the validity of the constructs and the narratives.

The appropriateness of the application of this methodology in my study is further warranted for the following reasons. First, its holistic nature is consistent with my research purpose. "The goal of comparative analysis is to determine the combinations of causal conditions that differentiate sets of cases. . . . Causal conditions are not examined separately, as in studies of covariation, but in combinations" (Ragin 1994: 116–19). Each case is understood "as a combination of causal conditions linked to a particular outcome" (Ragin 1994: 119). Second, this method is multilevel by nature, as it provides for data analysis at

two levels simultaneously—the system and within the system. It allows the use of system-level variables to explain variation across systems and within system relationships (Ragin 1987). For instance, in his study of international variations in class voting behavior, Alford (1963) used the degree of industrialization and urbanization to explain differences among countries in within-system relationships, such as the strength of the relationship between social classes and party support (Ragin 1987: 4). Hence, Alford's observational unit was the individual, and his explanatory unit—the society. Third, the method uses data that, in the final analysis, are presented in binary codes (for example, a condition is present or absent; is represented as of high or low degree). This allowed me to make full use of the qualitative and quantitative data I collected, since I was able to collapse existing categories and to develop new ones if the analysis dictated so. Fourth, the method stimulates "a rich dialogue between ideas and evidence," which allowed me to constantly check and recheck the validity of the constructs and the identified causal conditions. Last, the notions of sampling and sampling distributions are not relevant to the use of comparative method. As Ragin explains, it does not matter how many cases you have in the present or absent categories, since they do not need to be equally distributed (1987: 52).

The data analysis proceeds in several sequential steps. The first concerns an examination of the data relevant to the independent causes. My study has two major components: factors in relation to the projects' formal organizational structure, and factors pertaining to the projects' social networks structure. All of the quantitative data pertinent to the factors measuring the projects' formal organization were entered into an SPSS data file, checked for consistency and congruency, and "cleaned." I then examined all categories of interest and checked those against the respondents' narratives and the themes that emerged from the interviews. Simultaneously, I developed project histories from the qualitative data in order to situate the data into the specific context. The second step in preparing the data for the QCA involves an aggregation of data. After I examined the descriptive statistics on the variables of interest (for example, categories, frequencies, average scores), I aggregated that data at the project level. This was followed by a data-reduction procedure. The result from the latter was the development of categories, which I codi-

fied into binary codes. For instance, the degree of projects' formal structure centralization was codified as "high" and "low" centralization; consultative leadership style was codified as a condition "present" or "absent"; and the degree of technical autonomy as "high" and "low." The aim of this step was to make the data set consistent with the requirements of the comparative analysis methodology.

I followed the same procedures with reference to the social networks data. I entered and processed these data with UCINET 6.0 software for relational data analysis (Borgatti, Everett, and Freeman 2002). After examining the results, I codified them into binary codes at the project and individual levels (for example, "high" and "low" degrees of projects' density and centralization, and "high" and "low" degrees of individual's centrality in each of the four social networks).

The ultimate goal of the QCA is to identify the configurations of conditions associated with the outcome of interest, in this case "high success" of technologically innovative projects. Thus, my next step was to enter the categorized data (reduced to binary codes) pertaining to the conditions associated with the projects' formal and social networks structures and to the outcomes into a data matrix table, on the basis of which a "truth table" was constructed.[5] The methodology then requires the analysis to continue by constantly comparing configurations of causes, that is, the rows of the constructed table, as opposed to simply comparing the presence or absence of conditions, in other words, comparing columns. This is to allow for the elimination of the noncausal conditions from the model, and, ultimately, to elicit the combinations of conditions conducive to successful projects' outcome (Ragin 1994, 1987). This is a process during which one starts with one or two variables, pursues the contradictions,[6] revises the independent variables (that is, eliminates the irrelevant ones and adds new ones), and continues so until all the contradictions are resolved and a set of relevant independent causes is identified. As a result of applying this algorithm, the following changes were introduced to the list of independent variables. The variables "degree of formal structure centralization," "leadership style," "degree of technical autonomy," "task interdependency," "network density," "network centralization," and "individual's centrality" in the expressive and instrumental networks were dropped

from the model. Their presence resulted in a large number of configurations with contradictory outcomes that, due to the insufficient number of empirical cases in my data set, could not be resolved in any other way but by their exclusion (Coverdill, Finlay, and Martin 1994; Kniss 1997; Ragin 1987). The examination of the interview material at this stage confirmed the relevance of the remaining three variables and, in addition, gave prominence to a new one that emerged from the interviews with the laboratory director and the project managers, namely, "degree and duration of corporate support for the project." Hence, I added the latter causal condition to the model.

I used a software package, QCA 3.1, developed by Ragin and Drass (1992), as a data-processing tool. I then checked the results and the findings against the narratives that I had developed out of the qualitative data. Finally, tests were run to evaluate the *necessity* and *sufficiency* of each causal combination, and a Boolean minimization technique was applied to the ones that passed successfully.

The sufficiency test is guided by the question, "How many positive cases and/or negative cases are enough to establish the sufficiency of a causal combination?" The answer to it is strongly influenced by the number of cases portraying the combination. Two types of sufficiency tests are used in answer to this question, *veristic* (with and without specified frequency criteria) and *probabilistic* (binomial test for $n = 30$ or less and z-test for $n > 30$). Probability tests, naturally, yield more certainty in the outcome. The appropriate probability test in this case is the binomial one, for my data set is less than 30. According to Ragin, the binomial test "assesses the probability of observing a specific range of 'successful' outcomes, given an expected probability of success, which in turn is provided by the benchmark selected by the investigator" (2000: 112). The three benchmark levels suggested by Ragin for testing causal combinations are .50 ("sufficient more often than not"), .65 ("usually sufficient"), and .85 ("almost always sufficient") at three levels of significance: $\alpha = .10$, $\alpha = .05$, and $\alpha = .01$. To use the .65 benchmark, the minimum required number of cases displaying the causal combination is seven (see Table 4.9 in Ragin 2000: 114), for benchmark of .85 it is eleven, and for .50 it is five. As my data set would not result in more than six cases that display any causal combination, the only appropriate benchmark and significance levels for my

analysis are respectively .50 and $\alpha = .05$. Table A2.3 shows the distribution of the combinations of causal conditions that have passed the sufficiency test for "high success." Fourteen combinations altogether have displayed the (S) outcome. Their probabilities were calculated based on the following binomial probability formula:

$$\binom{N}{r} p^r q^{N-r},$$

where:

N = the number of cases displaying the causal combination;

r = number of cases displaying the outcome;

p = benchmark proportion;

q = 1 − p

The observed proportions (.83 and 1) for the combinations that have passed the binomial sufficiency test are superior to the benchmark proportion (.50). Thus, I concluded that this is not by chance. The results indicate that for a benchmark of .50, "sufficient more often than not," and a significance level of .05, the probabilities (0.0312 and 0.0430) of observing five or more successful outcomes that display the same causal combinations are smaller than the significance level. This lends support to the claim that the causal combinations listed above are "sufficient more often than not" for an outcome of "high project success."

After running the QCA minimization procedure on all the possible combinations for which there was empirical support, the final solution that incorporates the necessary and the sufficient conditions is: S = f • C • T • O.[7] It reads as "a project that is characterized by a low degree of formal reporting communication (f) *and* a high degree of corporate support (C), *and* on which there is a person(s) who occupies a highly central position in the technical-advice network (technical competence) (T), *and* on which there is a person(s) who occupies a highly central position in the organizational-advice network (O), 'more often than not' will result in a 'high success' outcome." In other words, the result shows that these four critical success factors shape the success on technical

Table A2.3

Causal combinations that have passed the sufficiency test at .50 benchmark and a significance level of $\alpha = .05$

Combination Number	Row Number from the Truth Table	Grouping	Number of Cases (Projects) Displaying the Causal Combination	Proportion of Cases (Projects) Displaying the "High Success" Outcome (S)	Probability ($\alpha = .05$)
1	8	f•C•T•O	5	1	0.0312
2	18	f•C•T	5	1	0.0312
3	28	f•C•O	5	1	0.0312
4	36	f•T•O	5	1	0.0312
5	48	C•T•O	5	1	0.0312
6	50	f•C	5	1	0.0312
7	54	f•T	6	.83	0.0430
8	58	f•O	5	1	0.0312
9	64	C•T	5	1	0.0312
10	72	C•O	5	1	0.0312
11	73	f	6	.83	0.0430
12	76	C	5	1	0.0312
13	78	T	6	.83	0.0430
14	80	O	5	1	0.0312

projects in conjuncture, not separately. This result does not in any way suggest that this is the only combination of factors that is likely to enhance success on technical projects. With a higher number of projects and more variety among them, other factors and combinations are also likely to be confirmed.

Finally, I uncovered the unique way in which positions and roles are assigned at Global East's R&D laboratory by mapping the results from the qualitative analyses onto the results from the social networks analysis and then onto the positions and roles that respondents held on their projects. I checked the findings that resulted from the mapping procedure against the narratives of the laboratory members. This I discussed in more detail in Chapter 3.

The collection of qualitative and quantitative data, as well as the application of the three analytical methodologies—inductive analysis, the QCA, and the social network analysis—has proven rather illuminating to my findings. Not only was I able to describe each structure and to identify the overlaps between the *prescribed* and the *actual* patterns of connections and authority at all project's levels, but I was also able to understand the forces and the mechanisms that have produced them. Ultimately, employing a combination of qualitative and quantitative methodologies proved to be of utmost importance to the comprehension of the relationship and the dynamics between the projects' formal and informal structures: whether the two simply complement one another or are interwoven, and, if so, through what mechanism and with what effect on the outcome. As my findings show, understanding of the processes through which a prescribed relationship is turned into an actual one and vice versa is critical. In the absence of rich qualitative data this would have been impossible, just as an understanding of the glue that holds the system together would have been missed.

The application of the QCA, along with inductive and social network analyses, was instrumental to uncovering the specific conditions under which "stars" affect outcomes of projects positively. A reliance on the inductive analysis pointed to the importance of technical and organizational human capital, while a reliance on social network and inductive analyses identified the members occupying positions of centrality in the lab's social networks and revealed the source and significance of their social capital. It was, however, the application of the QCA that afforded me the discovery of the missing link.

Namely, it was the discovery that the human capital relevant to technological innovation—technical and organizational skills and knowledge—matters if and when supplanted by social capital that emanates from centrality in the two work-related advice networks.

More important, the existing literature on teams, small groups, and project management has demonstrated that none of the multitude of critical success factors that have been identified in prior research can shape the outcomes of technical projects on their own. In consequence, it is not simply the finding about the four specific conditions but their combination and the very nature of their relationship, as uncovered by the analysis of the qualitative data and confirmed through the application of the QCA, that adds new knowledge about the dynamics of R&D project management. In the absence of QCA, the interrelated nature of the effect of the four causal factors would have remained either suggestive (on the basis of the findings from the inductive analysis) or uncovered. What is more, as I noted earlier, I had initially excluded "corporate support for projects" from the list of independent conditions, as the findings from prior research on the effect of this variable have been inconsistent. It was only the analysis of the interview data and the built-in methodological flexibility of the QCA that permitted me to incorporate this causal condition in the final model. As the results suggest, its exclusion would have produced a gross omission of a critical explanatory factor.

Finally, the simultaneous analysis of the qualitative and the social networks data allowed me not only to describe the social structures in the R&D laboratory but to understand *why* and *how* the two work-related advice networks emerged in the first place, as well as the specific ways in which centrality in each contributed to the shaping of the projects' outcomes.

ISSUES OF VALIDITY AND RELIABILITY

Validity

I took the following steps to increase the construct validity of the concepts that I rely on. Whenever possible, I employed operational measures of the concepts, which had been used in other researchers' survey studies and published in peer-reviewed academic journals. In addition, during the interviews, I asked the respondents to elaborate on their answers and to provide specific examples.

Thus, I was able to check the meaning of each concept during the interview process. For instance, when discussing the degree of formal reporting to superiors on the project, I asked for the subject's understanding and definition of the concepts "formal reporting" and frequency, and for examples pertinent to the project under discussion. I also asked respondents to compare their definitions and meanings across projects, whenever possible. To avoid idiosyncratic variation, I formulated multiple questions to measure the same concept in the questionnaires. To address the issue of the reliability of the subjects' memory, which may be suspect from collecting information based on past activities, I studied instances of innovation that were in the final stages of being completed or were recently completed. In addition, multiple sources of information were used (questionnaires, interviews, observation, and company's documents), the aim of which was to allow for the development of converging lines of inquiry, thus addressing the issue of construct validity, and to maintain a chain of evidence to address the reliability issue.

Reliability

To assess the reliability and to standardize the interview protocols, the questionnaires, and the procedures, my colleagues on the research team and I conducted a pilot study in a separate R&D laboratory of another *Fortune* 500 company. The consistency of the data collection was enhanced by the use of two interviewers only, which diminishes the possibility of introducing incompatible interviewing styles. Eighty-five percent of the interviews were conducted by me; the remaining 15 percent by the second interviewer, who is also a co-principal investigator on the study. To account for inconsistencies, I developed converging (or diverging) storylines by asking overlapping questions of the respondents who occupy different positions in the projects' formal structures (for example, a project manager, a researcher, a technician, and so on) and play different roles on each project.

PROTECTION OF HUMAN SUBJECTS FROM HARM

Babbie defines harm to research subjects as "emotional or psychological distress, as well as physical harm" (1992: 471). A number of steps were taken to address any potential for harming or causing discomfort to the participants

in this study. The research design, the procedures, and the interview protocols and questionnaires were made available to the Institutional Review Board (IRB) of Boston University for review and approval. The data-collection forms, the interview protocols, and the questionnaires were made available to the laboratory director and reviewed by him for sensitive or threatening questions. Although he made no suggestions for removing questions, he offered new categories to two questions, which he believed to be more reflective of the organizational structure and the nature of the research and development process. The first suggestion concerned categories to the question "Which of the suggested categories describes best your position on the project?" The second one concerned categories that explored areas of technical and financial discretion. I incorporated his suggestions in the final version of the data-collection forms.

The confidentiality rights of the participating company are protected by an agreement between the home organization of the research team and the funding organization, the National Science Foundation. According to it, neither the company nor the organizational members will be identified in any document, report, or form, whether written or verbal. All respondents were assigned numerical codes, accessible by the principal investigator of the research team and myself only. The codes, rather than names, were used on all the data-collection forms, and pseudonyms are used in any report of the results, including in this book. Information that is revealing of the company, the R&D laboratory, and its product lines was omitted. In addition, all participants signed a consent form, approved by the IRB of Boston University, prior to being interviewed or administered a questionnaire. The form explained the purpose of the study and the measures taken for the protection of the identity of the subjects.

NOTES

CHAPTER 1

1. Griffin reported in 1997 that almost all companies conducting scientific research and innovation make use of teams and teamwork.

2. The terms *work group* and *teams* have been used interchangeably lately in the literature on organizations (Guzzo and Dickson 1996).

3. Thomas (1994: 3) has offered an elegant argument as to how to reconcile the two diverging perspectives on the question of what drives organizational change—technological determinist and social choice. He proposed a theory of "how social *and* technical systems are jointly responsible for organizational structuring and change."

4. A summary of the six projects that formed the basis of the investigation and an in-depth discussion of the research methodology are provided in the two appendices.

CHAPTER 2

1. Access to the company was secured as part of a project funded by the National Science Foundation (NSF), NSF grant MOTI-9714058.

2. The terms *laboratory, lab,* and *facility* are used interchangeably.

3. The criteria used for selecting the projects are discussed in Appendix 2.

4. In addition, the project managers were also asked to classify the outcomes of the projects that they managed and to elaborate on their categorization. Although each respondent's narrative pointed to a multitude of success factors, there were no discrepancies in labeling the projects' outcomes.

5. The issue with including "true failures" at the time of the study was that these were aborted projects that never got to the point of even being assigned project members. Consequently, this renders the collection of social networks data and the examination of the effects of social relations and positions of centrality in both the formal and informal structure on these projects' outcomes impossible.

6. I also asked the laboratory director and the projects' managers to classify the outcomes of each of the six projects as an instance of either radical or incremental technological innovation. Unfortunately, I had to drop this way of classifying the projects from further analysis. I provide a detailed account for this in Appendix 2.

7. One respondent did not give his permission to tape-record the interview session.

8. For a detailed explanation of the methodology behind the study, please refer to Appendix 2.

9. For a breakdown by project, please see Table A2.2.

10. I followed Strauss and Corbin's (1990) and Lincoln and Cuba's (1985) strategy of inductive analysis. I offer a more detailed account of how I arrived at these four themes in Appendix 2.

11. These two projects were not part of the study, as neither one of them met the criteria for project selection. For instance, the project in question took place seven years prior to the data collection; including it in the investigations would have entailed serious problems with the accuracy of respondents' recollection and the construction of the social networks.

12. Please refer to Appendix 2 for a detailed account of this method, its application, and the analytical advantages that resulted from it. For further in-depth explanation on how the QCA method was applied to this data set, also see Rizova (2006b).

13. Please note that in logical statements, the operator "o" should read as logical *and* while the operator "+" means logical *or*.

CHAPTER 3

1. Consistent with both the realist and nominalist approaches of boundary definition (Knoke and Kuklinski 1982; Laumann, Marsden, and Prensky 1983), I used the R&D lab as the boundary of the social network.

2. Following Freeman (1979). The social network data were processed using UCINET 6 software (Borgatti, Everett, and Freeman 2002).

3. Specifically, as the data in the table show, it is between 1.53 and 1.65 standard deviations above the mean, or between the 87th and 90th percentiles.

4. The networks are drawn in KrackPlot 3.0 (Krackhardt, Blythe, and McGrath 1994).

5. Unfortunately, as I noted in Chapter 2, it was not possible to draw a clear distinction along these lines on the six projects that I investigated. A comparison of the staffing practices on *Beta, Gamma,* and *Delta,* on the one hand, and on *Epsilon,* on the other, suggests though that the nature of technical complexity of a project is taken into an account when assigning roles and responsibilities on projects.

6. Ibarra defines personal sources of power as "Expertise stemming from individual attributes such as experience, seniority, education, and professional activity . . ." (1993: 474).

7. These are discussed in Chapter 5. Social capital is embedded in, and emanates from, social relations (Burt 2000; Coleman 1988; Lin 2001).

8. The argument for the critical importance of culture in technology organizations has been convincingly made already. One of the most notable works on the matter is Gideon Kunda's 1992 ethnographic study of the engineering culture at *Tech.* In fact, if it was not for the size and age of his company, and a few other major differences in specific practices, there were times when I had the acute suspicion that we were researching the same

company. The management at Global East faces the same perpetual tensions documented by Kunda between organizing for creativity and the need for control. The engineering mentality, the informal social stratification system, the work ethic, and the daily challenges that I detected at Global East are very much the same ones uncovered and analyzed in Kunda's work. There were some notable differences as well. Junior engineers at Global East do not "usually do lowly and boring work and engage in narrowly defined tasks" (p. 40). The distinctions between social categories (engineers, managers, and technicians) are not pronounced in such a clear way at the lab as they are in *Tech*. The formal status between seniority among scientists and engineers is present, but this does not affect tasks and project assignments. The two very distinct professional tracks are not formally set out either. There are no formal presentations from the laboratory or the divisional management in the way it happens in *Tech*. Most important, while "culture" replaces "structure" as an organizing principle to explain reality and guide action at *Tech* (Kunda 1992: 30), at Global East it seems as though the structure creates and in turn is reinforced and supported, but not replaced, by an organizational culture.

CHAPTER 4

1. Ancona and Caldwell (1992) describe a similar situation with the "ID printer project."
2. Ancona and Caldwell (1992: 660), too, have found that "cycles may play a role in team behavior," and hence, changing strategies may be needed to support high performance over time.

CHAPTER 5

1 The benchmark and significance levels are respectively .50 and $\alpha = .05$. For further methodological details, please refer to Appendix 2.
2. The idea for this separation came from my prior industrial experience and from insights from the pilot study for this project conducted in another *Fortune* 500 company in the Boston area.

APPENDIX 2

1. In accordance with the confidentiality agreement, I do not identify the company, its products, the projects under investigation, or the respondents. Any names used in the book are pseudonyms.
2. The four social networks are discussed in more detail in Chapter 2.
3. Narratives are the ways in which "respondents in interviews impose order on the flow of experience to make sense of events and actions in their lives" (Riessman 1993: 2).
4. The QCA method has been widely used in comparative-historical and political research (Brown and Boswell 1995; Griffin, Botsko, Wahl, and Isaac 1991; Janoski and Hicks 1994). There, it has been employed to carry out comparative studies of the causes and outcomes of social revolutions (Skocpol 1979; Wickham-Crowley 1991); to investigate the effect of human capital and state intervention on the performance of health care systems (Hollingsworth, Hanneman, Hage, and Ragin 1996); to examine *how* and *why* social networks influence religious conversion (Smilde 2005); and to account for the variance in labor manage-

ment practices in manufacturing plants (Coverdill and Finlay 1995). In 1994, *Sociological Methods and Research* devoted a special issue to qualitative methods, and a number of articles concerning the application and advantages of the QCA have found a place in it.

5. The "truth table" lists all the possible combinations of present and absent conditions of the independent causes.

6. Contradictions are those configurations of independent variables that result in a mixed outcome (0 and 1).

7. Operator "o" in logical statement indicates logical *and*, as opposed to the operator "+", which indicates logical *or*.

REFERENCES

Adler, P., and S. Kwon. 2002. "Social Capital: Prospects for a New Concept." *Academy of Management Review, 27*, 17–40.

Ahuja, G. 2000. "Collaboration Network, Structural Holes, and Innovation: A Longitudinal Study." *Administrative Science Quarterly, 45*, 425–455.

Alford, R. 1963. *Party and Society.* Chicago: University of Chicago Press.

Allen, T. 1977. *Managing the Flow of Technology.* Cambridge, MA: MIT Press.

———. 1984. *Managing the Flow of Technology: Technology Transfer and the Dissemination of Technological Information Within the R&D Organization.* Cambridge, MA: MIT Press.

Allen, T., M. Tushman, and D. Lee. 1979. "Technology Transfer as a Function of Position in the Spectrum from Research Through Development to Technical Services." *Academy of Management Journal, 22*, 694–708.

Amabile, T. M. 1988. "A Model of Creativity and Innovation in Organizations." In B. Staw and L. Cummings (eds.), *Research in Organizational Behavior,* vol. 10, 123–167. Greenwich, CT: JAI Press.

Amenta, E., and J. Poulsen. 1994. "Where to Begin: A Survey of Five Approaches to Selecting Independent Variables for Qualitative Comparative Analysis." *Sociological Methods and Research, 23*(1), 22–53.

Ancona, D. 1990. "Outward Bound: Strategies for Team Survival in an Organization." *Academy of Management Journal, 33*, 334–365.

Ancona, D., and D. F. Caldwell. 1992. "Bridging the Boundary: External Activity and Performance in Organizational Teams." *Administrative Science Quarterly, 37*, 634–665.

Ancona, D., H. Bresman, and K. Kaeufer. 2002. "The Comparative Advantage of X-Teams." *MIT Sloan Management Review, 43,* 33–39.

Anderson, N., and N. King. 1993. "Innovation in Organizations." In C. L. Cooper and I. T. Robertson (eds.), *International Review of Industrial and Organizational Psychology,* vol. 8, 1–34. New York: Wiley & Sons.

Argote, L., and R. Ophir. 2002. "Intraorganizational Learning." In J. Baum (ed.), *Companion to Organizations,* 181–207. Oxford: Blackwell Business.

Argote, L., S. Beckman, and D. Epple. 1990. "The Persistence and Transfer of Learning in Industrial Settings." *Management Science, 36,* 140–154.

Argote, L., B. McEvily, and R. Reagans. 2003. "Managing Knowledge in Organizations: Creating, Retaining, and Transferring Knowledge." *Management Science, 49*(4), 571–582.

Argote, L., P. Ingram, J. Levine, and R. Moreland. 2000. "Knowledge Transfer in Organizations." *Organizational Behavior and Human Decision Processes, 82,* 1–8.

Arrow, H., J. E. McGrath, and J. L. Berdahl. 2000. *Small Groups as Complex Systems: Formation, Coordination, Development, and Adaptation.* London: Sage.

Astebro, T. 2004. "Key Success Factors for Technological Entrepreneurs' R&D Projects." *IEEE Transactions on Engineering Management, 51*(3), 314–321.

Ayers, D. J., G. L. Gordon, and D. D. Schoenbachler. 2001. "Integration and New Product Development Success: The Role of Formal and Informal Controls." *The Journal of Applied Business Research, 17,* 133–148.

Babbie, E. 1992. *The Practice of Social Research.* Belmont, CA: Wadsworth.

Balachandra, R., and J. H. Friar. 1997. "Factors for Success in R&D Projects and New Product Innovation: A Contextual Framework." *IEEE Transactions on Engineering Management, 44,* 3, 276–287.

———. 1999. "Managing New Product Development Processes the Right Way." *Information, Knowledge, & Systems Management, 1,* 33–43.

Balkundi, P., and D. Harrison. 2006. "Ties, Leaders, and Time in Teams: Strong Inference About Network Structure's Effect on Team Viability and Performance." *Academy of Management Journal, 49,* 49–68.

Barley, S. R. 1990. "An Alignment of Technology and Structure Through Roles and Networks." *Administrative Science Quarterly, 35,* 61–103.

Barley, S. R., and G. Kunda. 2004. *Gurus, Hired Guns, and Warm Bodies: Itinerant Experts in a Knowledge Economy.* Princeton and Oxford: Princeton University Press.

Baum, J. C., and C. Oliver. 1992. "Institutional Embeddedness and the Dynamics of Organizational Populations." *American Sociological Review, 57,* 540–559.

Becker, G. 1964. *Human Capital.* New York: Columbia University Press.

Blau, J. R., and R. Alba. 1982. "Empowering Nets of Participation." *Administrative Science Quarterly, 27*(3), 363–379.

Blau, P. M. 1955. *The Dynamics of Bureaucracy.* Chicago: University of Chicago Press.

Blau, P. M., and R. Schoenherr. 1971. *The Structure of Organizations.* New York: Basic Books.

Blau, P. M., and W. R. Scott. 1962. *Formal Organizations: A Comparative Approach.* San Francisco: Chandler.

Bohnet, I. 1997. *Cooperation and Communication.* Munich, Germany: Mohr/Siebeck.

Borgatti, S., and R. Cross. 2003. "A Relational View of Information Seeking and Learning in Social Networks." *Management Science, 49*(4): 432–445.

Borgatti, S., M. G. Everett, and L. C. Freeman. 2002. *UCINET 6 for Windows.* Cambridge, MA: Analytic Technologies.

Bourdieu, P. 1986. "The Forms of Capital." In J. G. Richardson (ed.), *Handbook of Theory and Research for the Sociology of Education,* 241–258. New York: Greenwood Press.

Brass, D. J. 1984. "Being in the Right Place: A Structural Analysis of Individual Influence in an Organization." *Administrative Science Quarterly, 29,* 518–539.

Brass, D. J., and M. Burkhardt. 1993. "Potential Power and Power Use: An Investigation of Structure and Behavior." *Academy of Management Journal, 36,* 441–470.

Brass, D. J., and G. Labianca. 1999. "Social Capital, Social Liabilities, and Social Resources Management." In T.A.J. Leenders and S. M. Gabbay (eds.), *Corporate Social Capital and Liability,* 323–338. Boston: Kluwer Academic Publishers.

Brass, D. J., J. Galaskiewicz, H. R. Greve, and W. Tsai. 2004. "Taking Stock of Networks and Organizations: A Multilevel Perspective." *Academy of Management Journal, 47,* 795–817.

Brint, S. 2001. "Professionals and the 'Knowledge Economy': Rethinking the Theory of Postindustrial Society." *Current Sociology, 49*(4), 101–132.

Brown, C., and R. Boswell. 1995. "Strikebreaking or Solidarity in the Great Steel Strike of 1919: A Split Labor Market, Game-Theoretic, and QCA Analysis." *American Journal of Sociology, 100,* 1479–1519.

Brown, S., and K. Eisenhardt. 1995. "Product Development: Past Research, Present Findings, and Future Directions." *Academy of Management Review, 20*(2), 343–378.

———. 1997. "The Art of Continuous Change: Linking Complexity Theory and Time-Paced Evolution in Relentlessly Shifting Organizations." *Administrative Science Quarterly, 42*, 1–34.

———. 1998. *Competing on the Edge: Strategy as Structured Chaos.* Boston: Harvard Business School Press.

Burningham, C., and M. A. West. 1995. "Individual, Climate, and Group Interaction Processes as Predictors of Work Team Innovation." *Small Group Research, 26*, 106–117.

Burns, T., and G. Stalker. 1961. *The Management of Innovation.* London: Tavistock Publications.

Burt, R. S. 1983. "Distinguishing Relational Contents." In R. S. Burt and M. J. Minor (eds.), *Applied Network Analysis.* Beverly Hills, CA: Sage.

———. 1980. "Models of Network Structure." *Annual Review of Sociology, 6*, 79–141.

———. 1992. *Structural Holes.* Cambridge, MA: Harvard University Press.

———. 2000. "The Network Structure of Social Capital." In R. Sutton and B. M. Staw (eds.), *Research in Organizational Behavior,* vol. 22, 345–423. Greenwich, CT: JAI Press.

Canner, N., and N. J. Mass. 2005. "Turn R&D Upside Down." *Research Technology Management, 48*, 17–21.

Carr, A. 1996. *Managing the Change Process: A Field Book for Change Agent Consultants.* London: Publishing Division, Coopers & Lybrand.

Castellacci, F., S. Grodal, S. Mendonca, and M. Wibe. 2005. "Advances and Challenges in Innovation Studies." *Journal of Economic Issues, 39*, 91–121.

Chandler, A. 1962. *Strategy and Structure: Chapters in the History of the American Industrial Enterprise.* Cambridge, MA.: MIT Press.

Clark, K., and T. Fujimoto. 1991. *Product Development Performance.* Boston: Harvard Business School Press.

Coleman, J. S. 1988. "Social Capital in the Creation of Human Capital." *American Journal of Sociology, 94* (Supplement), 95–120.

———. 1990. *Foundations of Social Theory.* Cambridge, MA: Harvard University Press.

Cooper, R. 1979. "The Dimensions of Industrial New Product Success and Failure." *Journal of Marketing, 43*, 93–103.

Cooper, R., and E. Kleinschmidt. 1987. "New Products: What Separates Winners from Losers." *Journal of Product Innovation Management, 4,* 169–184.

Coverdill, J., and W. Finlay. 1995. "Understanding Mills via Mill-Type Methods: An Application of Qualitative Comparative Analysis to a Study of Labor Management in Southern Textile Manufacturing." *Qualitative Sociology, 18,* 457–478.

Coverdill, J., W. Finlay, and J. Martin. 1994. "Labor Management in the Southern Textile Industry." *Sociological Methods and Research, 23*(1), 54–85.

Cozijnsen, A. J., W. J. Vrakking, and M. van Ijzerloo. 2000. "Success and Failure of 50 Innovation Projects in Dutch Companies." *European Journal of Innovation Management, 3,* 150–159.

Cross, R., and L. Sproull. 2004. "More Than an Answer: Information Relationships for Actionable Knowledge." *Organization Science, 15,* 446–460.

Cross, R., S. Borgatti, and A. Parker. 2002. "Making Invisible Work Visible: Using Social Networks Analysis to Support Strategic Collaboration." *California Management Review, 44,* 25–46.

Crossan, M., H. Lane, and R. White. 1999. "An Organizational Learning Framework: From Intuition to Institution." *Academy of Management Review, 24,* 522–537.

Cummings, J. N. 2004. "Work Groups, Structural Diversity, and Knowledge Sharing in a Global Organization." *Management Science, 50,* 352–364.

Dalton, M. 1959. *Men Who Manage.* New York: Wiley & Sons.

Damanpour, F. 1988. "Innovation Type, Radicalness, and the Adoption Process." *Communication Research, 15,* 545–567.

———. 1991. "Organizational Innovation: A Meta-Analysis of Effects, Determinants and Moderators." *Academy of Management Journal, 34,* 555–590.

Dobrev, S., T. Kim, and L. Solari. 2004. "The Two Sides of the Coin: Core Competence as Capability and Obsolescence." In J. Baum and A. McGahan (eds.), *Advances in Strategic Management,* vol. 21. Greenwich, CT: JAI Press.

Dougherty, D. 2001. "Re-Imagining the Differentiation and Integration of Work for Sustained Product Innovation." *Organization Science, 12,* 612–631.

Drazin, R., and C. Schoonhoven. 1996. "Community, Population, and Organization Effects on Innovation: A Multilevel Perspective." *Academy of Management Journal, 39,* 1065–1084.

Dvir, D., T. Raz, and A. Shenhar. 2003. "An Empirical Analysis of the Relationship Between Project Planning and Project Success." *International Journal of Project Management, 21,* 89–95.

Ebadi, Y., and J. Utterback. 1984. "The Effects of Communication on Technological Innovation." *Management Science, 30,* 572–585.

Ederer, P. 2006. *Innovation at Work: The European Human Capital Index.* Lisbon Council. Available at *www.lisboncouncil.net.*

Einhorn, H. 1972. "Expert Measurement and Mechanical Combination." *Organizational Behavior and Human Performance, 7,* 86–106.

Espy, S. N. 1986. *Handbook of Strategic Planning for Nonprofit Organizations.* New York: Praeger.

Fagerberg, J. 2005. "Innovation: A Guide to the Literature." In J. Fagerberg, D. Mowery, and R. Nelson (eds.), *The Oxford Handbook of Innovation,* 1–26. Oxford: Oxford University Press.

Fiol, C., and M. Lyles. 1985. "Organizational Learning." *Academy of Management Journal, 10*(4), 803–813.

Fleming, L., and O. Sorenson. 2000. "Technology as a Complex Adaptive System: Evidence from Patent Data." *Research Policy 30,* 1019–1039.

Freeman, C., A. Robertson, B. Achilladelis, and P. Jervis. 1972. "Success and Failure in Industrial Innovation, Report on Project SAPPHO by the Science Policy Research Unit." London: Center for the Study of Industrial Innovation, University of Sussex.

Freeman, L. C. 1979. "Centrality in Social Networks: Conceptual Clarification." *Social Networks, 1,* 215–239.

Fukuyama, F. 1995. *Trust: The Social Virtues and the Creation of Prosperity.* New York: Free Press.

Galbraith, J. 1973. *Designing Complex Organizations.* Reading, MA: Addison-Wesley.

Gargiulo, M. 1993. "Two-Step Leverage: Managing Constraint in Organizational Politics." *Administrative Science Quarterly, 39,* 1–19.

Gargiulo, M., and M. Benassi. 1999. "The Dark Side of Social Capital." In T.A.J. Leenders and S. M. Gabbay (eds.), *Corporate Social Capital and Liability,* 298–321. Boston: Kluwer Academic Publishers.

Gibbons, D. 2004. "Friendship and Advice Networks in the Context of Changing Professional Values." *Administrative Science Quarterly, 49,* 238–262.

Giddens, A. 1979. *Central Problems in Social Theory: Action, Structure and Contradiction in Social Action.* Berkeley: University of California Press.

Gnyawali, D. 2001. "Cooperative Networks and Competitive Dynamics: A Structural Embeddedness Perspective." *The Academy of Management Review, 26,* 431–445.

Grabher, G. 1993. "The Weakness of Strong Ties: The Lock-In of Regional Development in the Ruhr Area. In G. Grabher (ed.), *The Embedded Firm: On the Socioeconomics of Industrial Networks,* 255–277. New York: Routledge.

Granovetter, M. 1985. "Economic Action and Social Structure: The Problem of Embeddedness." *American Journal of Sociology, 91,*3, 481–510.

———. 1995. *Getting a Job: A Study of Contacts and Careers,* 2nd ed. Chicago and London: University of Chicago Press.

Green, S., M. Gavin, and L. Aiman-Smith. 1995. "Assessing a Multidimensional Measure of Radical Technological Innovation. *IEEE Transactions on Engineering Management, 42,* 203–214.

Griffin, A. 1997. *Drivers of NPD Success: The 1997 PDMA Report.* University of Illinois.

Griffin, A., and A. Page. 1996. "PDMA Success Measurement Project: Recommended Measures for Product Development Success and Failure." *Journal of Product Innovation Management, 13,* 478–496.

Griffin, L. J., C. Botsko, A. Wahl, and L. Isaac. 1991. "Theoretical Generality, Case Particularity: Qualitative Comparative Analysis of Trade Union Growth and Decline." *International Journal of Comparative Sociology, XXXII,* 1–2, 110–136.

Gulati, R., and M. Gargiulo. 1999. "Where Do Networks Come From?" *American Journal of Sociology, 104,* 1439–1493.

Guzzo, R., and M. Dickson. 1996. "Teams in Organizations: Recent Research on Performance and Effectiveness." *Annual Review of Psychology, 46,* 307–338.

Guzzo, R., and G. P. Shea. 1992. "Group Performance and Intergroup Relations in Organizations." In M. D. Dunnette and L. M. Hough (eds.), *Handbook of Industrial and Organizational Psychology,* vol. 3, 269–313. Palo Alto, CA: Consulting Psychologists Press.

Hackman, R. J. 1987. "The Design of Work Teams." In J. W. Lorsch (ed.), *Handbook of Organizational Behavior,* 315–342. Englewood Cliffs, NJ: Prentice-Hall.

Hage, J. T. 1980. *Theories of Organizations.* New York: Wiley & Sons.

Hage, J. T., and M. Aiken. 1970. *Social Change in Complex Organizations.* New York: Random House.

Hage, J.T., and R. Hollingsworth. 2000. "A Strategy for the Analysis of Idea Innovation Network and Institutions." *Organization Studies, 21,* 971–1004.

Hansen, M. T. 1999. "The Search-Transfer Problem: The Role of Weak Ties in Sharing Knowledge Across Organization Subunits." *Administrative Science Quarterly, 44,* 82–111.

————. 2002. "Knowledge Networks: Explaining Effective Knowledge Sharing in Multiunit Companies." *Organization Science, 13,* 232–248.

Hansen, M., M. Mors, and B. Løvås. 2005. "Knowledge Sharing in Organizations: Multiple Networks, Multiple Phases." *Academy of Management Journal, 48*(5): 776–793.

Hargadon, A. B., and R. I. Sutton. 1997. "Technology Brokering and Innovation in a Product Development Firm." *Administrative Science Quarterly, 42,* 716–749.

Higgins, M., and N. Nohria. 1999. "The Sidekick Effect: Mentoring Relationships and the Development of Social Capital." In T.A.J. Leenders and S. M. Gabbay (eds.), *Corporate Social Capital and Liability,* 161–179. Boston: Kluwer Academic Publishers.

Hindle, T. 2006. "The New Organisation," *McKinsey Quarterly;* quoted in "A Survey of the Company," *The Economist,* January 21, *378*(8461), 3–5.

Hollingsworth, R., R. Hanneman, G. Hage, and C. Ragin. 1996. "The Effect of Human Capital and State Intervention on the Performance of Medical Systems." *Social Forces, 75,* 459–484.

Ibarra, H. 1993. "Network Centrality, Power, and Innovation Involvement: Determinants of Technical and Administrative Roles." *Academy of Management Journal, 36,* 471–501.

IBM Center for the Business of Government. *Research Announcement 2006–2007.* Available at *www.businessofgovernment.org.*

Ilgen, D. R., J. R. Hollenbeck, M. Johnson, and D. Jundt. 2005. "Teams in Organizations: From Input-Process-Output Models to IMOI Models." *Annual Review of Psychology, 56,* 17–43.

Ingram, P., and P. Roberts. 2000. "Friendship Among Competitors in the Sydney Hotel Industry." *American Journal of Sociology, 196,* 387–423.

Janoski, T., and A. Hicks. 1994. *The Comparative Political Economy of the Welfare State.* Cambridge, MA: Cambridge University Press.

Johnson, D. A., and J. D. Linton. 2000. "Social Networks and the Implementation of Environmental Technology." *IEEE Transactions on Engineering Management, 47,* 465–477.

Katz, R., and T. Allen. 1985. "Project Performance and the Locus of Influence in the R&D Matrix." *Academy of Management Journal, 28,* 67–87.

Katz, R., and M. Tushman. 1981. "An Investigation into the Managerial Roles and Career Paths of Gatekeepers and Project Supervisors in a Major R&D Facility." *R&D Management, 11,* 103–110.

Keller, R. 1994. "Technology-Information Processing Fit and the Performance of R&D Project Groups: A Test of Contingency Theory." *Academy of Management Journal, 37,* 167–179.

———. 2001. "Cross-Functional Project Groups in Research and New Product Development: Diversity, Communications, Job Stress, and Outcomes." *Academy of Management Journal, 44*(3), 547–555.

Kerssens-Van Drongelen, I. C., P. C. Weerd-Nederhof, and O.A.M. Fisscher. 1996. "Describing the Issues of Knowledge Management in R&D: Towards a Communication and Analysis Tool." *R&D Management, 26,* 213–230.

Klein, K. J., H. Tosi, and A. A. Cannella Jr. 1999. "Multilevel Theory Building: Benefits, Barriers, and New Developments." *Academy of Management Review, 24,* 243–248.

Kleinman, D., and S. Vallas. 2001. "Science, Capitalism, and the Rise of the 'Knowledge Worker': The Changing Structure of Knowledge Production in the United States." *Theory and Society, 30,* 451–492.

Kleinschmidt, E., and R. Cooper. 1995. "The Relative Importance of New Product Success Determinants: Perception Versus Reality." *R&D Management, 25,* 282–297.

Kniss, F. 1997. *Disquiet in the Land: Cultural Conflict in the American Mennonite Communities.* New Brunswick, NJ, and London: Rutgers University Press.

Knoke, D. 1999. "Organizational Networks and Corporate Social Capital." In R.T.A. Leenders and S. M. Gabbay (eds.), *Corporate Social Capital and Liability,* 17–42. Boston: Kluwer Academic Publishers.

Knoke, D., and J. Kuklinski. 1982. *Network Analysis.* Newbury Park, CA; London; and New Delhi: Sage.

Kostova, T. 1999. "Transnational Transfer of Strategic Organizational Practices: A Contextual Perspective." *Academy of Management Review, 24,* 308–324.

Krackhardt, D. 1990. "Assessing the Political Landscape: Structure, Cognition, and Power in Organizations." *Administrative Science Quarterly, 35,* 342–369.

Krackhardt, D., and J. R. Hanson. 1993. "Informal Networks: The Company Behind the Chart." *Harvard Business Review, 71,* 104–111.

Krackhardt, D., J. Blythe, and C. McGrath. 1994. "KrackPlot 3.0: An Improved Network Drawing Program." *Connections, 17,* 53–55.

Kunda, G. 1992. *Engineering Culture: Control and Commitment in a High-Tech Corporation.* Temple University Press.

Labianca, G., D. Brass, and B. Gray. 1998. "Social Networks and Perceptions of Inter-Group Conflict: The Role of Negative Relationships and Third Parties." *Academy of Management Journal, 41,* 55–67.

Lam, A. 2005. "Organizational Innovation." In J. Fagerberg, D. Mowery, and R. Nelson (eds.), *The Oxford Handbook of Innovation,* 115–147. Oxford: Oxford University Press.

Laumann, E. O., P. V. Marsden, and D. Prensky. 1983. "The Boundary Specification Problem in Network Analysis." In R. S. Burt and M. J. Minor (eds.), *Applied Network Analysis,* Beverly Hills, CA: Sage.

Lawrence, P., and J. Lorsch. 1967. "Differentiation and Integration in Complex Organizations." *Administrative Science Quarterly, 12,* 1–47.

Levitt, B., and J. March. 1988. "Organizational Learning." *Annual Review of Sociology, 14,* 319–340.

Liebeskind, J. P., A. L. Oliver, L. Zuker, and M. Brewer. 1996. "Social Networks, Learning, and Flexibility: Sourcing Scientific Knowledge in New Biotechnology Firms." *Organization Science, 7,* 428–443.

Lin, N. 2001. *Social Capital: A Theory of Social Structure and Action.* New York: Cambridge University Press.

Lincoln, Y., and E. Cuba. 1985. *Naturalistic Inquiry.* Newbury Park, CA: Sage.

Lindenberg, S. 1996. "Constitutionalism Versus Rationalism: Two Views of Rational Choice Sociology." In J. Clark (ed.), *James S. Coleman,* 229–311. London: Falmer Press.

Maidique, M., and B. J. Zirger. 1984. "A Study of Successes and Failure in Product Innovation: The Case of the U.S. Electronics Industry." *IEEE Transactions on Engineering Management, 31,* 192–203.

March, J. 1991. "Exploration and Exploitation in Organizational Learning." *Organization Science, 2,* 71–87.

March, J., and H. Simon. 1958. *Organizations.* New York: Wiley & Sons.

Marsden, P. V. 1990. "Network Data and Measurement." *Annual Review of Sociology, 16,* 435–463.

McGrath, J. 1984. *Groups: Interaction and Performance.* Englewood Cliffs, NJ: Prentice-Hall.

McGrath, J., and L. Argote. 2001. "Group Processes in Organizational Contexts." In M. A. Hogg and R. S. Tindale (eds.), *Blackwell Handbook of Social Psychology, Vol.3: Group Processes,* 603–627. Oxford: Blackwell.

McGrath, J., H. Arrow, and J. Berdahl. 2000. "The Study of Groups: Past, Present, and Future." *Personality and Social Psychology Review, 4,* 95–105.

Mehra, A., M. Kilduff, and D. Brass. 2001. "The Social Networks of High and Low Self-Monitors: Implications for Workplace Performance." *Administrative Science Quarterly, 46,* 121–146.

Merton, R. 1938. "Science and the Social Order." *Philosophy of Science, 5*(3), 321–337.

———. 1957. *Social Theory and Social Structure.* New York: Free Press.

Meyer, M., and J. Utterback. 1995. "Product Development Cycle Time and Commercial Success." *IEEE Transactions on Engineering Management, 42*(4), 297–304.

Mintzberg, H. 1979. *The Structuring of Organization.* Englewood Cliffs, NJ: Prentice-Hall.

Monge, P., and N. Contractor. 2001. "Emergence of Communication Networks." In F. Jablin and L. Putnam (eds.), *The New Handbook of Organizational Communication,* 440–502. Thousand Oaks, CA: Sage.

Mote, J. 2005. "R&D Ecology: Using 2-Mode Network Analysis to Explore Complexity in R&D Environments." *Journal of Engineering and Technology Management, 22,* 93–111.

Nahapiet, J., and S. Ghoshal. 1998. "Social Capital, Intellectual Capital, and the Organizational Advantage." *Academy of Management Review, 23,* 242–266.

Nebus, J. 2006. "Building Collegial Information Networks: A Theory of Advice Network Generation." *Academy of Management Review, 31,* 615–637.

Nee, V. 1998. "Norms and Networks in Economic and Organizational Performance." *The American Economic Review, 88,* 85–89.

Nohria, N., and R. Gulati, 1994. "Firms and Their Environments." In N. J. Smelser and R. Swedberg (eds.), *The Handbook of Economic Sociology,* 529–551. Princeton, NJ: Princeton University Press.

Oh, H., G. Labianca, and M. Chung. 2006. "A Multilevel Model of Group Social Capital." *Academy of Management Review, 31,* 569–583.

Oliver, A. L., and J. P. Liebeskind. 1998. "Three Levels of Networking for Sourcing Intellectual Capital in Biotechnology: Implications for Studying Interorganizational Networks." *International Studies of Management & Organization, 27,* 76–103.

Olson, E., O. Walker, and R. Ruekert. 1995. "Organizing for Effective New Product Development: The Moderating Role of Product Innovativeness." *Journal of Marketing, 59,* 48–62.

Page, A. 1993. "Assessing New Product Development Practices and Performance: Establishing Crucial Norms." *Journal of Product Innovation Management, 10,* 273–287.

Patrashkova, R., and S. Comb. 2004. "Exploring Why More Communication Is Not Better: Insights From a Computational Model of Cross-functional Teams." *Journal of Engineering and Technology Management, 21,* 83–114.

Peter, L., and R. Hull. 1969. *The Peter Principle.* New York: William Morrow.

Perrow, C. 1967. "A Framework for the Comparative Analysis of Organizations." *American Sociological Review, 32,* 194–208.

Pfeffer, J., and G. Salancik. 1978. *The External Control of Organizations: A Resource Dependence Perspective.* New York: Harper and Row.

Pinto, J., and D. Slevin. 1988. "Project Success: Definitions and Measurement Techniques." *Project Management Journal, 19:* 67–73.

Podolny, J., and J. Baron. 1997. "Resources and Relationships: Social Networks and Mobility in the Workplace." *American Sociological Review, 62,* 673–693.

Podolny, J., and K. Page. 1998. "Network Forms of Organization." *Annual Review of Sociology, 24,* 57–76.

Portes, A., and J. Sensenbrenner. 1993. "Embeddedness and Immigration: Notes on the Social Determinants of Economic Action." *American Journal of Sociology, 98*(6), 1320–1350.

Powell, W., and S. Grodal. 2005. "Networks of Innovators." In J. Fagerberg, D. Mowery, and R. Nelson (eds.), *The Oxford Handbook of Innovation,* 56–85. Oxford: Oxford University Press.

Powell, W., K. Koput, and L. Smith-Doerr. 1996. "Interorganizational Collaboration and the Locus of Innovation: Networks of Learning in Biotechnology." *Administrative Science Quarterly, 41,* 116–145.

Powell, W., and K. Snellman. 2004. "The Knowledge Economy." *Annual Review of Sociology, 30,* 199–220.

Powell, W. W. 1998. "Learning from Collaboration: Knowledge and Networks in Biotechnology and Pharmaceutical Industries." *California Management Review, 40,* 228–240.

Powell, W. W., and P. Brantley. 1992. "Competitive Cooperation in Biotechnology: Learning Through Networks?" In N. Nohria and R. Eccles (eds.), *Networks and Organizations: Structure, Form and Action,* 366–394. Boston: Harvard Business School.

Pugh, D. S., D. J. Hickson, C. R. Hinings, and C. Turner. 1968. "Dimensions of Organization Structure." *Administrative Science Quarterly, 13,* 115–126.

Putnam, R. 1993. *Making Democracy Work: Civic Traditions in Modern Italy.* Princeton, NJ: Princeton University Press.

———. 1995. "Bowling Alone: America's Declining Social Capital." *Journal of Democracy, 6,* 5–78.

Quinn, J. B. 2000. "Outsourcing Innovation: The New Engine for Growth." *Sloan Management Review, 41*(4), 13–28.

Ragin, C. 1987. *The Comparative Method: Moving Beyond Qualitative and Quantitative Strategies.* Berkeley and Los Angeles, and London: University of California Press.

———. 1994. *Constructing Social Research.* Thousand Oaks, CA; London; and New Delhi: Pine Forge Press.

———. 2000. *Fuzzy-Set Social Science.* Chicago and London: University of Chicago Press.

Ragin, C., and K. Drass. 1992. QCA 3.1. software. Copyright 1992–1998 Kriss A. Drass.

Reagans, R., and E. Z. Zuckerman. 2001. "Networks, Diversity, and Productivity: The Social Capital of Corporate R&D Teams." *Organization Science, 12,* 502–517.

Reagans, R., E. Zuckerman, and B. McEvily. 2004. "How to Make the Team: Social Networks vs. Demography as Criteria for Designing Effective Teams." *Administrative Science Quarterly, 49,* 101–133.

Riessman, C. 1993. *Narrative Analysis.* Thousand Oaks, CA: Sage.

Rivkin, J., and N. Siggelkow. 2003. "Balancing Search and Stability: Interdependencies Among Elements of Organization Design." *Management Science, 49,* 290–311.

Rizova, P. S. 2002. *The Secret of Success: A Study of Six Technologically Innovative Projects at a Research and Development Laboratory.* Ph.D. dissertation, Boston University.

———. 2006(a). "Are You Networked for Successful Innovation?" *MIT Sloan Management Review,* Spring, *47,* 49–55.

———. 2006(b). "Applying Charles Ragin's Method of Qualitative Comparative Analysis (QCA) to the Study of Technological Innovation." Presentation at the Academy of Management Meeting, Atlanta, August 11–16.

Roberts, K., and C. O'Reilly. 1979. "Some Correlates of Communication Roles in Organizations." *Academy of Management Journal, 22,* 42–57.

Rothwell, R. 1992. "Successful Industrial Innovation: Critical Success Factors for the 1990s." *R&D Management, 3,* 221–239.

Rothwell, R., et al. 1974. "SAPPHO Updated-Project SAPPHO Phase II." *Research Policy, 3,* 258–291.

Rubenstein, A., A. Chakrabarti, R. O'Keefe, W. Souder, and H. Young. 1976. "Factors Influencing Innovation Success at the Project Level." *Research Management, 19,* 15–19.

Rulke, D., and J. Galaskiewicz. 2000. "Distribution of Knowledge, Group Network Structure, and Group Performance." *Management Science, 46*(5): 612–625.

Saxenian, A. 1988. "The Cheshire Cat's Grin: Innovation and Regional Development in England." *Technology Review,* Feb.-Mar., 67–75.

———. 1996. *Regional Advantage: Culture and Competition in Silicon Valley and Route 128.* Cambridge, MA: Harvard University Press.

Schumpeter, J. 1934. *The Theory of Economic Development.* Cambridge, MA: Harvard University Press.

Scott, J. 2005. *Social Network Analysis: A Handbook,* 2nd ed. London: Sage.

Scott, S. G. 1997. "Social Identification Effects in Product and Process Development Teams." *Journal of Engineering Technology Management, 14,* 97–127.

Scott, W. R. 1992. *Organizations: Rational, Natural, and Open Systems.* Englewood Cliffs, NJ: Prentice-Hall.

Segal. A. 2004. "Is America Losing Its Edge?" *Foreign Affairs,* Nov.-Dec., *83*(6): 2–9.

Shan, W., G. Walker, and B. Kogut. 1994. "Interfirm Cooperation and Startup Innovation in the Biotechnology Industry." *Strategic Management Journal, 15,* 387–394.

Shaw, M. E. 1964. "Communication Networks." In L. Berkowitz (ed.), *Advances in Experimental Social Psychology,* 111–147. New York: Academic Press.

Shenhar, A. 2001. "One Size Does Not Fit All Projects: Exploring Classical Contingency Domains." *Management Science, 47,* 394–414.

Shrader, C., J. Lincoln, and A. Hoffman. 1989. "The Network Structures of Organizations: Effects of Task Contingencies and Distributional Form." *Human Relations, 42,* 43–66.

Simmel, G. 1902. "The Number of Members as Determining the Sociological Form of the Group." *American Journal of Sociology, 8,* 1–46.

Skocpol, T. 1979. *States and Social Revolutions: A Comparative Analysis of France, Russia, and China.* Cambridge, MA: Cambridge University Press.

Smilde, D. 2005. "A Qualitative Comparative Analysis of Conversion to Venezuelan Evangelicalism: How Networks Matter." *American Journal of Sociology, 111,* 757–796.

Smith-Doerr, L., and W. W. Powell. 2005. "Networks and Economic Life." In N. J. Smelser and R. Swedberg (eds.), *The Handbook of Economic Sociology*, 379–402. Princeton: Princeton University Press.

Smith-Doerr, L., I. Manev, and P. Rizova. 2004. "The Meaning of Success: Network Position and the Social Construction of Project Outcomes in an R&D Lab." *Journal of Engineering and Technology Management, 21*, 51–81.

Sparrowe, R., R. Liden, S. Wayne, and M. Kraimer. 2001. "Social Networks and the Performance of Individuals and Groups." *Academy of Management Journal, 44*, 316–325.

Starbuck, W., and P. Nystrom. 1981. "Designing and Understanding Organizations." In P. Nystrom and W. Starbuck (eds.), *Handbook of Organizational Design*, vol. 1. Oxford: Oxford University Press.

Starbuck, W., A. Greve, and B. Hedberg. 1978. "Responding to Crisis." *Journal of Business Administration, 9*(2), 112–137.

Stark, D. 1996. "Recombinant Property in East European Capitalism." *American Journal of Sociology, 101*, 993–1027.

Stevenson, W., and M. Gilly. 1991. "Information Processing and Problem Solving: The Migration of Problems Through Formal Positions and Networks of Ties." *Academy of Management Journal, 34*, 918–928.

Stewart, G. L. 2006. "A Meta-Analytic Review of Relationships Between Team Design Features and Team Performance." *Journal of Management, 32*, 29–54.

Stewart, G. L., and M. R. Barrick. 2000. "Team Structure and Performance: Assessing the Mediating Role of Intrateam Process and the Moderating Role of Task Type." *Academy of Management Journal, 43*, 135–148.

Strauss, A., and J. Corbin. 1990. *Basics of Qualitative Research: Grounded Theory Procedures and Techniques*. Newbury Park, CA: Sage.

"A Survey of the Company: The New Organization." *The Economist*, January 21, 2006, 3–18.

Takeuchi, H., and I. Nonaka. 1986. "The New New Product Development Game." *Harvard Business Review*, Jan.–Feb., 137–146.

Thomas, R. J. 1994. *What Machines Can't Do: Politics and Technology in the Industrial Enterprise*. Berkeley, Los Angeles, London: University of California Press.

Thompson, J. 1967. *Organizations in Action*. New York: McGraw-Hill.

Tichy, N. 1980. "Networks in Organizations." In P. C. Nystrom and W. Starbuck (eds.), *Handbook of Organizational Design*, vol. 2. London: Oxford University Press.

Tichy, N., M. Tushman, and C. Fombrun. 1979. "Social Network Analysis for Organizations." *Academy of Management Review, 4,* 507–519.

Tsai, W. 2001. "Knowledge Transfer in Intraorganizational Networks: Effects of Network Position and Absorptive Capacity on Business Unit Innovation and Performance." *Academy of Management Journal, 44,* 996–1004.

Tushman, M. 1977. "Communication Across Organizational Boundaries: Special Boundary Roles in the Innovation Process." *Administrative Science Quarterly, 22,* 587–605.

———. 1978. "Technical Communication in R&D Laboratories: The Impact of Project Work Characteristics." *Academy of Management Journal, 21,* 624–645.

Tushman, M., and D. Nadler. 1986. "Organizing for Innovation." *California Management Review,* Spring, *28,3,* 74–93.

Tushman, M., and T. Scanlan. 1981. "Boundary-Spanning Individuals: Their Role in Information Transfer and Their Antecedents." *Academy of Management Journal, 24,* 289–305.

Utterback, J. 1974. "Innovation in Industry and the Diffusion of Technology." *Science, 183,* 658–662.

———. 1994. *Mastering the Dynamics of Innovation: How Companies Can Seize Opportunities in the Face of Technological Change.* Boston: Harvard Business School Press.

Uzzi, Brian. 1996. "The Sources and Consequences of Embeddedness for the Economic Performance of Organizations: The Network Effect." *American Sociological Review, 61,* 674–698.

———. 1997. "Networks and the Paradox of Embeddedness." *Administrative Science Quarterly, 42,* 35–67.

———. 1999. "Embeddedness in the Making of Financial Capital: How Social Relations and Networks Benefit Firms Seeking Financing." *American Sociological Review, 64,* 481–505.

Vallas, S. P. 1999. "Rethinking Post-Fordism: The Meaning of Workplace Flexibility." *Sociological Theory, 17,* 68–101.

———. 2006. "Empowerment Redux: Structure, Agency, and the Remaking of Managerial Authority." *American Journal of Sociology, 111,* 1677–1717.

Van der Panne, G., C. van Beers, and A. Kleinknecht. 2003. "Success and Failure of Innovation: A Literature Review." *International Journal of Innovation Management, 7,* 309–338.

Van de Ven, A. 1986. "Central Problems in the Management of Innovation." *Management Science, 32,* 590–607.

Von Hippel, E. 1994. "'Sticky Information' and the Locus of Problem Solving: Implications for Innovation." *Management Science, 40*(4): 429–439.

Wasserman, S., and K. Faust. 1994. *Social Network Analysis: Methods and Applications.* Cambridge, UK: Cambridge University Press.

Wegner, D. 1995. "A Computer Network Model of Human Transactive Memory." *Social Cognition, 13,* 319–339.

West, M. 1990. "The Social Psychology of Innovation in Groups." In M. A. West and J. L. Farr (eds.), *Innovation and Creativity at Work: Psychological and Organizational Strategies.* Chichster, UK: Wiley & Sons.

West, M. A., G. Hirst, A. Richter, and H. Shipton. 2004. "Twelve Steps to Heaven: Successfully Managing Change Through Developing Innovative Teams." *European Journal of Work and Organizational Psychology, 13,* 269–299.

Wickham-Crowley, T. 1991. "A Qualitative Comparative Approach to Latin American Revolutions." *International Journal of Comparative Sociology, XXXII,* 1–2, 82–109.

Williamson, O. E. 1991. "Comparative Economic Organization: The Analysis of Discrete Structural Alternatives." *Administrative Science Quarterly, 36,* 269–296.

Woodward, J. 1958. *Management and Technology.* London: H.M.S.O.

INDEX